Springer Theses

Recognizing Outstanding Ph.D. Research

Aims and Scope

The series "Springer Theses" brings together a selection of the very best Ph.D. theses from around the world and across the physical sciences. Nominated and endorsed by two recognized specialists, each published volume has been selected for its scientific excellence and the high impact of its contents for the pertinent field of research. For greater accessibility to non-specialists, the published versions include an extended introduction, as well as a foreword by the student's supervisor explaining the special relevance of the work for the field. As a whole, the series will provide a valuable resource both for newcomers to the research fields described, and for other scientists seeking detailed background information on special questions. Finally, it provides an accredited documentation of the valuable contributions made by today's younger generation of scientists.

Theses are accepted into the series by invited nomination only and must fulfill all of the following criteria

- They must be written in good English.
- The topic should fall within the confines of Chemistry, Physics, Earth Sciences, Engineering and related interdisciplinary fields such as Materials, Nanoscience, Chemical Engineering, Complex Systems and Biophysics.
- The work reported in the thesis must represent a significant scientific advance.
- If the thesis includes previously published material, permission to reproduce this must be gained from the respective copyright holder.
- They must have been examined and passed during the 12 months prior to nomination.
- Each thesis should include a foreword by the supervisor outlining the significance of its content.
- The theses should have a clearly defined structure including an introduction accessible to scientists not expert in that particular field.

More information about this series at http://www.springer.com/series/8790

Alex Ganose

Atomic-Scale Insights into Emergent Photovoltaic Absorbers

Doctoral Thesis accepted by
the University College London, UK

 Springer

Author
Dr. Alex Ganose
Energy Technologies Area
Lawrence Berkeley National Laboratory
Berkeley, CA, USA

Supervisor
Prof. David O. Scanlon
Department of Chemistry
University College London
London, UK

ISSN 2190-5053   ISSN 2190-5061   (electronic)
Springer Theses
ISBN 978-3-030-55710-2   ISBN 978-3-030-55708-9   (eBook)
https://doi.org/10.1007/978-3-030-55708-9

This Springer imprint is published by the registered company Springer Nature Switzerland AG
The registered company address is: Gewerbestrasse 11, 6330 Cham, Switzerland

Supervisor's Foreword

Reduction of the global dependence on fossil fuels, and the realization of effective renewable energy technologies are much sought after goals in this environmentally conscious age. Construction of clean energy systems is, therefore, a key objective, and central to this is our need to develop more efficient materials for harnessing solar energy to power our increasingly energy hungry world. The current market for solar absorbers is dominated by Si and CdTe. Both of these materials, however, are not ideal for this purpose; Si possesses an indirect band gap, and therefore, is not an efficient absorber, and CdTe is comprised of toxic (Cd) and rare (Te) elements, and so is not ideal for wide scale deployment. The hunt is, therefore, on to find earth-abundant, efficient materials to challenge their dominance.

This thesis by Dr. Alex Ganose represents a tour de force of contemporary Computational Materials Chemistry, marrying rational design with sophisticated ab initio screening techniques to test the viability of a range of emerging solar absorbers. The thesis features an overview of how to carry out state-of-the-art calculations on PV materials, with the results chapters divided into two sections: (i) on perovskite-derived semiconductors, including two-dimensional perovskite pseudohalides and vacancy-ordered double perovskites, and (ii) on bismuth chalcoiodides. Where possible, the computational work is linked or compared with experiments, and even makes predictions to guide the design of devices based on our analysis. Notably, Alexs work on BiSI and BiSeI suggested that the previous solar cell device architectures were sub optimal, and proposed the use of poly (9,9-di-n-octylfluorenyl-2,7-diyl) (F8) as a more effective hole transporting layer which might increase device performance. This has subsequently been experimentally verified, with F8 being successfully used in the device with record efficiency (ACS Appl. Energy Mater., 2, 3878 (2019)). This illustrates the power of predictive modelling in the field of photovoltaics.

London, UK
April 2020

Prof. David O. Scanlon

Abstract

The world is currently experiencing an energy crisis; as energy reserves continue to be depleted at record pace, there is a growing demand for a clean and renewable energy source capable of sustaining economic growth. Arguably, solar power is the most promising renewable technology due to the enormous amount of energy that reaches the earth in the form of solar radiation. Traditional solar cells, such as those based on crystalline silicon, have achieved efficiencies up to 25% but are limited in their widespread application due to limits in their cost-competitiveness. Recently, the lead hybrid perovskites have emerged as a highly efficient class of solar absorber, with efficiencies reaching over 23% within just nine years. Unfortunately, the stability of these materials is poor and concerns over the toxicity of lead have sparked significant research effort toward the search for alternative absorbers capable of achieving comparable efficiencies.

This thesis investigates a number of materials for their suitability as solar absorbers. Throughout, ab initio methods are used to provide insight into the structural, electronic and optical properties that determine their performance. Special attention is paid to understanding the behaviour of intrinsic defects due to their critical role in determining carrier recombination and transport. Initially, perovskite-based materials are discussed, including a new family of layered perovskites, and the lead-free vacancy-ordered double perovskites. In the second part of this thesis, the search for emerging photovoltaics is extended to a promising family of bismuth-based absorbers, of interest due to their non-toxic and earth-abundant nature. Throughout this work, we aim to provide specific guidance for experimental researchers hoping to produce more efficient photovoltaic devices.

Publications

Parts of this thesis have been published in the following journal articles:

1. D. M. Fabian, A. M. Ganose, J. W. Ziller, D. O. Scanlon, M. C. Beard, and S. Ardo, Influence of One Specific CarbonCarbon Bond on the Quality, Stability, and Photovoltaic Performance of Hybrid OrganicInorganic Bismuth Iodide Materials, *ACS Applied Energy Materials*, **2**, 1579–1587 (2019)
2. M. M. S. Karim, A. M. Ganose, L. Pieters, W. W. Leung, J. Wade, M. Birkett, L. Zhang, D. O. Scanlon, and R. G. Palgrave, Anion Distribution, Structural Distortion, and Symmetry-Driven Optical Band Gap Bowing in Mixed Halide Cs2SnX6 Vacancy Ordered Double Perovskites, *Chemistry of Materials*, **31** 9430–9444 (2019)
3. A. E. Maughan, A. M. Ganose, D. O. Scanlon, and J. R. Neilson, Perspectives and design principles of vacancy-ordered double perovskite halide semiconductors, *Chemistry of Materials*, **31**, 1184–1195 (2019)
4. Z. Wang, A. M. Ganose, C. Niu, and D. O. Scanlon, Two-dimensional eclipsed arrangement hybrid perovskites for tunable energy level alignments and photovoltaics, *Journal of Materials Chemistry C*, **7** 5139–5147 (2019)
5. K. J. Fallon, P. Budden, E. Salvadori, A. M. Ganose, C. N. Savory, L. Eyre, D. O. Scanlon, C. W. M. Kay, A. Rao, R. H. Friend, A. J. Musser, and H. Bronstein, Exploiting excited-state aromaticity to design highly stable singlet fission materials, *Journal of the American Chemical Society*, **141**, 13867–13876 (2018)
6. A. M. Ganose, S. Matsumoto, J. Buckeridge, and D. O. Scanlon, Defect Engineering of Earth-Abundant Solar Absorbers BiSI and BiSeI, *Chemistry of Materials*, **30**, 3827–3835 (2018)
7. A. E. Maughan, A. M. Ganose, M. A. Almaker, D. O. Scanlon, and J. R. Neilson, Tolerance Factor and Cooperative Tilting Effects in Vacancy-Ordered Double Perovskite Halides, *Chemistry of Materials*, **30**, 3909–3919 (2018)
8. A. M. Ganose, A. J. Jackson, and D. O. Scanlon, sumo: Command-line tools for plotting and analysis of ab initio calculations, *Journal of Open Source Software*, **3**, 717 (2018)
9. Z. Wang, A. M. Ganose, C. Niu, and D. O. Scanlon, First-principles insights into tin-based two-dimensional hybrid halide perovskites for photovoltaics, *Journal of Materials Chemistry A*, **6**, 5652–5660 (2018)
10. A. E. Maughan, A. M. Ganose, A. M. Candia, J. T. Granger, D. O. Scanlon, and J. R. Neilson, Anharmonicity and Octahedral Tilting in Hybrid Vacancy-Ordered Double Perovskites, *Chemistry of Materials*, **30**,472–483 (2018)
11. C. N. Savory, A. M. Ganose, and D. O. Scanlon, Exploring the $PbS-Bi_2S_3$ series for next generation energy conversion materials, *Chemistry of Materials*, **29**, 5156–5167 (2017)

12. A. M. Ganose, C. N. Savory, and D. O. Scanlon, $(CH_3NH_3)_2PbI_2(SCN)_2$ analogues for photovoltaic applications, *Journal of Materials Chemistry A*, **5**, 7845–7853 (2017)

13. A. M. Ganose, C. N. Savory, and D. O. Scanlon, Beyond methylammonium lead iodide: prospects for the emergent field of ns^2 containing solar absorbers, *Chemical Communications*, **53**, 20–44 (2017)

14. C. N. Savory, A. M. Ganose, W. Travis, R. S. Atri, R. G. Palgrave, and D. O. Scanlon, An Assessment of Silver Copper Sulphides for Photovoltaic Applications: Theoretical and Experimental Insights, *Journal of Materials Chemistry A*, **4**, 12648–12657 (2016)

15. A. E. Maughan, A. M. Ganose, M. M. Bordelon, E. M. Miller, D. O. Scanlon, and J. R. Neilson, Defect Tolerance to Intolerance in the Vacancy Ordered Double Perovskite Semiconductors Cs_2SnI_6 and Cs_2TeI_6, *Journal of the American Chemical Society*, **138**, 8453–8464 (2016)

16. A. M. Ganose, M. Cuff, K. T. Butler, A. Walsh, and D. O. Scanlon, Interplay of Orbital and Relativistic Effects in Bismuth Oxyhalides: BiOF, BiOCl, BiOBr and BiOI, *Chemistry of Materials*, **28**, 1980 (2016)

17. W. Travis, C. Knapp, C. N. Savory, A. M. Ganose, P. Kafourou, X. Song, Z. Sharif, J. K. Cockcroft, D. O. Scanlon, H. Bronstein, and R. G. Palgrave, Hybrid Organic–Inorganic Coordination Complexes as Tunable Optical Response Materials, *Inorganic Chemistry*, **55**, 3393 (2016)

18. A. M. Ganose, K. T. Butler, A. Walsh, and D. O. Scanlon, Relativistic Electronic Structure and Band Alignment of BiSI and BiSeI: Candidate Photovoltaic Materials, *Journal of Materials Chemistry A*, **4**, 2060 (2016)

19. A. M. Ganose and D. O. Scanlon, Band gap and work function tailoring of SnO_2 for improved transparent conducting ability in photovoltaics, *Journal of Materials Chemistry C*, **4**, 1467 (2016)

20. A. M. Ganose, C. N. Savory, and D. O. Scanlon, $(CH_3NH_3)_2Pb(SCN)_2I_2$: a more stable structural motif for hybrid halide photovoltaics?, *Journal of Physical Chemistry Letters*, **6**, 4594 (2015)

Publications arising from work conducted during the course of this thesis but not directly related to it:

21. A. Regoutz, A. M. Ganose, L. Blumenthal, C. Schlueter, T.-L. Lee, G. Kieslich, A. K. Cheetham, G. M. Vinai, T. Pincelli, G. Panaccione, H. K. Zhang, R. G. Egdell, J. Lischner, D. O. Scanlon, and D. J. Payne, Insights into the electronic structure of OsO_2 using soft and hard x-ray photoelectron spectroscopy in combination with density functional theory, *Physical Review Materials*, **3**, 025001 (2018)

22. N. H. Bashian, S. Zhou, A. M. Ganose, J. W. Stiles, A. Ee, D. S. Ashby, M. Zuba, D. O. Scanlon, L. F. Piper, B. Dunn, and B. C. Melot, Correlated Polyhedral Rotations in the Absence of Polarons During Electrochemical Insertion of Lithium in ReO_3, *ACS Energy Letters*, **3** 2513–2519 (2018)

23. A. J. Jackson, A. M. Ganose, A. Regoutz, R. G. Egdell, and D. O. Scanlon, Galore: Broadening and weighting for simulation of photoelectron spectroscopy, *Journal of Open Source Software*, **3**, 773 (2018)

24. A. M. Ganose, L. Gannon, F. Fabrizi, H. Nowell, S. Barnett, H. Lei, X. Zhu, C. Petrovic, D. O. Scanlon, and M. Hoesch, Local corrugation and persistent charge density wave in $ZrTe_3$ with Ni intercalation *Physical Review B*, **97** 155103 (2018)

25. D. Biswas, A. M. Ganose, R. Yano, J. M. Riley, L. Bawden, O. J. Clark, J. Feng, L. Collins-Mcintyre, M. T. Sajjad, W. Meevasana, T. K. Kim, M. Hoesch, J. E. Rault, T. Sasagawa, D. O. Scanlon, and P. D. C. King, Narrow-band anisotropic electronic structure of ReS_2, *Physical Review B*, **96**, 085205 (2017)

26. C. H. Hendon, K. T. Butler, A. M. Ganose, D. O. Scanlon, G. A. Ozin, and A. Walsh, Electroactive Nanoporous Metal Oxides and Chalcogenides, *Journal of Materials Chemistry A*, **29**, 3663–3670 (2017)

27. N. F. Quackenbush, H. Paik, M. J. Wahila, S. Sallis, M. E. Holtz, X. Huang, A. M. Ganose, B. J. Morgan, D. O. Scanlon, Y. Gu, F. Xue, L.-Q. Chen, G. E. Sterbinsky, C. Schlueter, T.-L. Lee, J. C. Woicik, J.-H. Guo, J. D. Brock, D. A. Muller, D. A. Arena, D. G. Schlom, and L. F. J. Piper, Stability of the M2 phase of vanadium dioxide induced by coherent epitaxial strain, *Physical Review B*, **94**, 085105 (2016)

28. C. I. Hiley, D. O. Scanlon, A. A. Sokol, S. M. Woodley, A. M. Ganose, S. Angio, J. M. De Teresa, P. Manuel, D. D. Khalyavin, M. Walker, M. R. Lees, and R. I. Walton, Antiferromagnetism at T > 500K in the layered hexagonal ruthenate $SrRu_2O_6$, *Physical Review B*, **92**, 104413 (2015)

29. Y. Hu, N. Goodeal, Y. Chen, A. M. Ganose, R. G. Palgrave, H. Bronstein and M. O. Blunt, Probing the Chemical Structure of Monolayer Covalent-Organic Frameworks Grown via Schiff-Base Condensation Reactions, *Chemical Communications*, **52**, 9941–9944 (2016)

Acknowledgements

I owe a huge debt of gratitude to my supervisor Dr. David O. Scanlon—initially for taking the risk of hiring an organic chemist, and subsequently for his generous advice and support throughout the last four years. I feel incredibly fortunate to have had the opportunity of his mentorship. I would like to thank all members of the Scanlon Materials Theory Group. In particular, special thanks go to Chris Savory for being the font of all learning, Dr. Ben Williamson for his discerning taste in figures, Dr. Katie Inzani for her knowledge of pre-1990s music and Dr. Adam Jackson for many in-depth discussions. Additional thanks go to the long-lost members of KLB room 349, including Dougal, Ian, Mia and Rachel, for the constant supply of beer and conversation.

Professor Aron Walsh and many of his current/ex-group members, including Jarvist, John, Keith, Dan, Chris, Suzy and Lucy have been excellent fun at conferences and never stopped answering my Slack questions. Particular thanks go to Prof. Jamie Neilson and Annalise Maughan at CSU for making my visit so enjoyable. Professor Brent Melot, Prof. Louis Piper, Dr. Anna Regoutz and Dr. Ben Morgan deserve thanks for the number of drinks they have bought me at conferences.

This work was funded by EPSRC and Diamond Light Source through the EPSRC Centre for Doctoral Training in Molecular Modelling and Materials Science (EP/L015862/1). The computational work in this thesis was performed on a number of supercomputing centres: ARCHER UK National Supercomputing Service, via membership of the UK's HEC Materials Chemistry Consortium, funded by EPSRC (EP/L000202); Iridis cluster, provided by the EPSRC funded Centre for Innovation (EP/K000144/1 and EP/K000136/1); and the UCL Legion and Grace HPC facilities.

On a personal note, Nikul and Jack mean everything to me. Many thanks to DK and Standi in their own ways. Lastly, I would like to thank my family. My brothers, Ryan and Liam, for always being so sarcastic and keeping me in my place. My father for his constant support. And my mother for everything she has done for me.

Contents

Acronyms

Theory

DFT	Density functional theory
DFPT	Density functional perturbation theory
HF	Hartree–Fock
KS	Kohn–Sham
SCF	Self-consistent field
SOC	Spin–orbit coupling

Density Functionals

GGA	Generalised gradient approximation
HSE06	Heyd–Scuseria–Ernzerhof functional
LDA	Local density approximation
PBE	Perdew–Burke–Ernzerhof functional
PBEsol	Perdew–Burke–Ernzerhof functional revised for solid
PBE0	Perdew–Burke–Ernzerhof hybrid functional

Methods

D3	Grimme's D3 correction
PAW	Projector augmented wave
XRD	X-ray diffraction
VASP	Vienna Ab initio Simulation Package

Electronic Properties

CB   Conduction band
CBM   Conduction band minimum
DOS   Density of states
EA   Electron affinity
IP   Ionization potential
VB   Valence band
VBM   Valence band maximum

Photovoltaic Properties

AM 1.5G   Air mass 1.5 global spectrum
EQE   External quantum efficiency
FF   Fill factor
SRH   Shockley–Read–Hall

Materials

Ch   Chalcogenide
FA   Formammidinium
FTO   Fluorine-doped tin oxide
HTM   Hole transporting material
ITO   Tin-doped indium oxide
MA   Methylammonium
MAPI $CH_3NH_3PbI_3$
MAPSI $(CH_3NH_3)_2Pb_2I_2(SCN)_2$
P3HT   Poly(3-hexylthiophene)
TCO   Transparent conducting oxide
X   Halide

Notation

Notation

$a()$	Function
$a[\,]$	Functional
$\{a\}$	Set of all a
$\hat{a}$	Operator
a	Vector
Re	Real part

Functions

δ	Dirac delta function
Θ	Heaviside step function

Constants

Z	Atomic number
h	Reduced Planck constant
e	Electron charge
k_B	Boltzmann constant
m_e	Electron rest mass

Coordinates

r	Cartesian coordinate
R	Lattice
G	Reciprocal lattice

x   Generalised coordinate
k   Wave vector

Energies

E_F Fermi level
E_H Hartree energy
E_g Band gap
E_{xc} Exchange–correlation energy
T Kinetic energy

Operators

$\widehat{H}$ Hamiltonian operator
$\widehat{h}$ Single-particle Hamiltonian operator
$\widehat{T}$ Kinetic energy operator
∇^2 Laplace operator

Potentials

V_H Hartree potential
V_{ext} External potential
V_{xc} Exchange–correlation potential

Thermodynamics

μ Chemical potential
T Temperature
$\Delta_f H$ Formation energy

Electronic Properties

q Charge
g Density of states
ε_r Dielectric constant
α Madelung constant
μ Mobility
V_{oc} Open-circuit voltage
J_{sc} Short-circuit current
m^* Effective mass

Wavefunctions

Ψ Many-body wavefunction
χ Single-particle wavefunction
ψ Kohn–Sham wavefunction
ϕ Atomic-like orbital
ρ Electron density

List of Figures

List of Tables

Part I
Introduction

Chapter 1
Photovoltaics

The use of fossil fuels as an energy source has enabled the rapid industrialisation seen across the world since the 18th century. Today, the burning of oil, coal, and natural gas is at a record high and shows little sign of abating. Indeed, global energy consumption is expected to rise to over 30PW by 2050 from just 19PW in 2012 [1]. This inexorable demand for energy is problematic for several reasons: despite the relatively low cost of fossil fuels, their use is not sustainable due to finite reserves. Furthermore, and perhaps most pressing, the combustion of fossil fuels is the leading human source of greenhouses gases, which are detrimental both to the environment—where they are a major factor contributing towards global warming [2]—and to the health of the population—where one in eight of total global deaths are linked to high levels of air pollution [3]. Lastly, the uneven distribution of oil reserves creates significant challenges for nations without access to their own energy pipelines. Here, the European Union particularly suffers, with recent estimates suggesting that by 2020 over 80% of its energy supply will originate from outside its borders [4].

There is, therefore, a growing demand for a clean energy source capable of providing sustainable economic growth [5]. Of the currently available technologies, photovoltaics are perhaps the most promising due to the enormous amount of energy the sun can provide. Over 1,500 exawatt-hours of energy reach the earth in the form of solar radiation each year, dwarfing the total know reserves of oil, gas, and coal, which amount to fewer than 9 exawatt-hours [6, 7]. While current photovoltaic technologies have reached efficiencies high enough to capture a significant proportion of this energy, the cost of manufacturing these devices is a major roadblock preventing their widespread adoption. As such, there is a need for cost-effective solar cell technologies if photovoltaics are to compete in utility scale power generation [8].

© The Editor(s) (if applicable) and The Author(s), under exclusive license
to Springer Nature Switzerland AG 2020
A. Ganose, *Atomic-Scale Insights into Emergent Photovoltaic Absorbers*,
Springer Theses, https://doi.org/10.1007/978-3-030-55708-9_1

1.1   The Photovoltaic Effect

The capacity of certain materials to produce a voltage and electric current upon exposure to light has been known for almost 200 years. It was first documented in 1839 by Edmond Becquerel, who noted a light-generated electric current between two platinum electrodes placed in a solution of sodium chloride. Subsequent experiments by Adams and Day in 1877 confirmed this effect arose from the properties of light rather than heat,[1] prompting them to term this phenomenon the *photoelectric effect*— today, referred to as the *photovoltaic effect*.

The origins of the photovoltaic effect in a semiconductor can be summarised as follows: the absorption of light results in the photoexcitation of an electron from the occupied valence band to the unoccupied conduction band, producing an electron–hole pair. If the pair is able to separate in space (termed exciton dissociation), both the electron and hole may behave as free charge carriers and contribute to the conductivity of the material by providing a photocurrent. Furthermore, as the electron and hole possess opposing charges and are separated in energy by the band gap of the semiconductor, the resulting potential difference produces a photovoltage. The existence of both a current and a voltage therefore allows for the generation of useable electrical power.

1.2   Principles of Photovoltaic Devices

The three main challenges required to efficiently harness the energy produced by the photovoltaic effect are:

1. To maximise the absorption of light and generate a large photocurrent.
2. To maximise the potential difference between electron and hole and generate a large photovoltage.
3. To separate the electron–hole pair with high efficiency.

As we shall see later on, challenges 1 and 2 are interdependent and therefore must balanced against each other. The challenge presented by 3 is largely tackled through the formation of a *p–n* junction, as used in a traditional solid-state solar cell.

A *p–n* junction is formed when a *p*-type material (doped with excess holes) and an *n*-type material (doped with excess electrons) are brought into contact (Fig. 1.1). At the interface, some of the excess electrons in the *n*-type layer will diffuse into the *p*-type layer. Similarly, some of the excess holes in the *p*-type layer will diffuse into the *n-type* material. In the *n-type* layer, the diffusion of electrons leaves behind positively charged dopant atoms which are fixed in place in the crystal structure.

---

[1] Alongside the direct photoexcitation of electrons, a photovoltage can also arise due to light generated heating of a material. The resulting increase in temperature can produce a temperature gradient, which may create a voltage through the Seebeck effect. This mechanism is exploited in thermoelectrics—an emerging class of renewable energy materials.

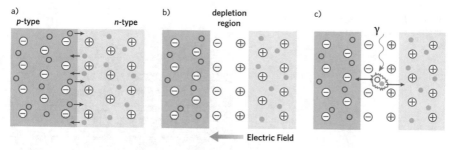

Fig. 1.1 Simplified representation of a *p–n* junction: **a** Upon contact of *p*-type (shaded blue) and *n*-type (shaded orange) semiconductors, majority carriers (orange circles for electrons; empty blue circles for holes) diffuse across the interface. **b** The diffusion leaves dopant ion cores exposed (negative ions in the *p*-type layer; positive ions in the *n*-type layer), leading to a build-up of opposing charges and a corresponding electric field at the interface. The region in which the electric field acts is termed the depletion region, since any free charge carriers will be quickly swept out. **c** Photoexcitation results in the formation of an electron–hole pair. The field acts to separate the exciton, with the hole extracted to the *p*-type layer and electron extracted to the *n*-type layer

Conversely, the diffusion of holes out of the *p*-type layer leaves behind negatively charged dopant sites. The resulting build-up of opposing charges at either side of the interface results in an electric field, which extends part way into the *n*- and *p*-type layers. The area in which this field acts is termed the depletion region, since any charge carriers generated in this region are quickly swept out by the electric field—holes toward to the *p*-type layer and electrons toward the *n*-type layer. As such, the region itself is largely depleted of free carriers. As exciton dissociation will be considerably more efficient if photoexcitation occurs within the depletion region, the doped semiconductors comprising the *p–n* junction are generally strong visible light absorbers.

To obtain useable electrical energy from a photovoltaic device, the separated charge carriers must be extracted out of the *p–n* junction and into an external circuit. Solar cells, therefore, require highly conductive front and back contacts to enable the transport of electrons and holes out of the device. Traditionally, a metal is used as the back contact for collection of holes. The front contact poses an additional challenge in that it must be transparent to allow light into the device. As such, transparent conducting oxides, such as indium tin oxide (ITO) or SnO_2, are employed as the *n*-type front contact layer. Many devices also contain buffer layers, which facilitate the transport of charge carriers to the electrodes. A simplified solar cell architecture is presented in Fig. 1.2.

1.2.1 Key Performance Characteristics of Solar Cells

The performance of solar cells is quantified by the power conversion efficiency, η, which measures the percentage of power output, P_{max}, to power input, P_{in}, according to

Fig. 1.2 Schematic of a
simplified *p–n* junction solar
cell

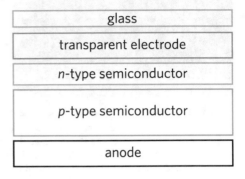

$$\eta = \frac{P_{\text{max}}}{P_{\text{in}}} = \frac{V_{\text{oc}} J_{\text{sc}} FF}{P_{\text{in}}}, \tag{1.1}$$

where V_{oc} is the open-circuit voltage, J_{sc} is the short-circuit current, and FF is
the fill factor. When measuring the efficiency of a photovoltaic, a set of standard
conditions are employed to ensure consistency across different devices. In particular,
measurements are performed at a temperature of 25 °C, using the AM 1.5G spectrum[2]
as an illumination source, and with the incident power defined as 1kW cm^{-2}.

To optimise the efficiency of a solar cell it is essential to maximise the terms
that contribute to the power output. The short-circuit current is the current passing
through the cell under illumination when the contacts on either side of the cell are
connected directly—resulting in zero voltage across the cell. The short-circuit current
is strongly dependent on the ability of the absorbing layer to generate charge carriers.
In an ideal device, every incident photon would be absorbed to produce an electron
and hole that can contribute to the overall current. As only photons with energies
greater than the band gap of the absorber can be absorbed, the largest short-circuit
current will be achieved when then band gap is smallest.

The open-circuit voltage is the maximum voltage across the cell under illumination if the net current is zero—i.e. when the external circuit is not complete. The
open-circuit voltage is greatest when there is the largest potential difference between
the separated electrons and holes. As such, the open-circuit voltage is controlled
primarily by the band gap of the semiconductor, with large band gaps allowing
for greater potential difference between the charge carriers. The theoretical maximum open-circuit voltage of a device is roughly equivalent to the band gap of the
absorber—assuming ohmic contacts with all buffer and transport layers. In practice,

---

[2]The AM 1.5G spectrum is a standard terrestrial reference spectra defined by the American Society
for Testing and Materials. The spectra is based on the global average spectral irradiance if the Sun
were at an angle of 42° and the solar cell were at sea level, including losses due water, ozone, and
other molecules in the atmosphere.

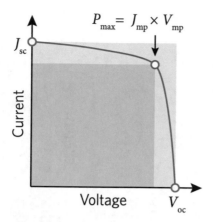

Fig. 1.3 Example J–V curve for a solar cell under illumination indicated by dark blue line, with the open-circuit voltage (V_{oc}) and short-circuit current (J_{sc}) highlighted. The dark blue square is defined by the location of the maximum power output (P_{max}), with the corresponding voltage and current at these points termed V_{mp} and J_{mp}. The fill factor is the ratio between the dark and light blue squares

however, the open-circuit voltage is reduced by radiative and non-radiative carrier recombination, which act to increase the forward bias diffusion current and limit device performance.

The final performance characteristic required to calculate the power conversion efficiency is the fill factor, which is determined from the current–voltage (J–V) curve of a cell under illumination. An example J–V curve for a hypothetical cell is provided in Fig. 1.3. The fill factor is a measure of the "squareness" of the J–V curve and is defined as the ratio between the maximum power output versus the product of V_{oc} and J_{sc}, according to

$$FF = \frac{P_{max}}{V_{oc}J_{sc}} = \frac{V_{mp}J_{mp}}{V_{oc}J_{sc}}, \tag{1.2}$$

where V_{mp} and J_{mp} are the voltage and current at the point of maximum power output. A fill factor closer to 1 indicates a highly ideal cell. In practice, recombination, interfacial effects, and other parasitic resistances generally result in a J–V curve that deviates significantly from the perfect behaviour.

1.2.2 Loss Mechanisms in Photovoltaic Devices

The performance of a solar cell is strongly impacted by device losses that can limit overall efficiency. When designing new absorber materials, it is crucial that these loss mechanisms are understood so their effects can be minimised. In solar absorbers, the dominating loss mechanisms are recombination of charge carriers, optical losses,

Fig. 1.4 Example plot of external quantum efficiency (EQE) against photon wavelength, indicated by a solid blue line. Regions of the spectrum are highlighted to demonstrate potential optical loss mechanisms

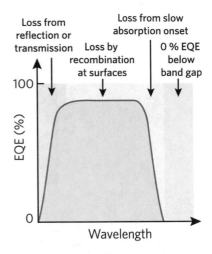

and parasitic resistances that occur across the cell. Parasitic resistances are generally introduced during device fabrication and have minimal dependence on the intrinsic properties of the absorber or contact layers. As such, this thesis is mainly concerned with optical and recombination losses.

Optical losses manifest as reduction in short-circuit current and arise due to incomplete conversion of incident photons into separated electron–hole pairs. The degree of optical loss in a device is quantified by the external quantum efficiency (EQE)—namely, the ratio of the number of carriers collected by the solar cell versus the total number of photons incident on the device.

The main sources of optical losses are transmission or reflection of visible light (Fig. 1.4). In particular, photons with energy below the band gap of the absorber will not be absorbed, resulting in an EQE of 0% for this energy range. Additional losses can result from front and rear surface reflection, and shading by the device electrodes.

Charge-carrier recombination is the process by which a meta-stable conduction band electron de-excites by combining with a hole in the valence band, thereby returning to equilibrium conditions. Recombination is detrimental for solar cell performance—by recombining, the free charge carriers cannot contribute the photocurrent (fewer charge carriers) and limit the photovoltage (increased dark forward bias current). The three most common types of electron–hole recombination are presented in Fig. 1.5 and are defined as:

Radiative Recombination. Also termed spontaneous emission, here an excited electron combines with a hole, releasing a photon in the process. Radiative recombination can be considered the reverse of absorption and is the most common mode of recombination in direct band gap semiconductors.

Shockley–Read–Hall (SRH) Recombination. All materials contain some concentration of defects (such as point defects, complexes, and grain boundaries), which can

Fig. 1.5 Representation of common electron–hole recombination mechanisms in photovoltaic devices: **a** Radiative recombination, **b** Shockley–Read–Hall recombination, and **c** Auger Recombination. Valence and conduction bands indicated by blue and orange lines, respectively. Holes and electrons indicated by empty blue circles and filled orange circles, respectively

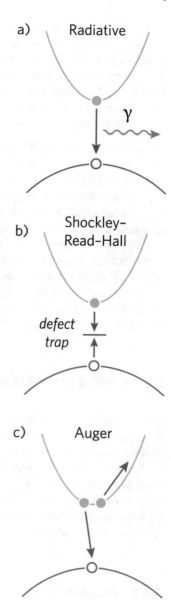

lead to the presence of isolated electronic states in the band gap. SRH or trap-assisted recombination is a two step process in which an electron (or hole) is trapped by the defect state, before a hole (or electron) is also trapped and the carriers recombine. If the trap is close in energy to the band edges (termed a shallow defect), thermal energy may be sufficient to promote the electron or hole back into the band where it can again act as a free carrier. As such, traps further away from the band edges, termed deep traps, are considerably more active as recombination centres.

Auger Recombination. Here, an electron and hole recombine, however, rather than emitting a photon or phonon, the energy is given to an electron in the conduction band, promoting it to a further excited state. This electron will then thermalise to back to the conduction band edge. As this is a three-particle process, the probability of Auger recombination is significantly lower than for the two-particle recombination pathways, except in heavily doped systems.

In practice, the rates of radiative and Auger recombination are largely independent of materials synthesis and device engineering methods. In contrast, Shockley–Read–Hall recombination depends on the type and number of defects in the system, so may be controlled by doping, surface passivation, and synthesis technique. As such, attempting to tune the rate of Shockley–Read–Hall recombination is a primary strategy when optimising a potential photovoltaic material.

1.3   Desired Solar Absorber Properties

It is difficult to predict how a novel solar absorber will perform in practice, due to the dependence of efficiency on many external conditions, including the deposition method, quality of precursors, and device architecture. However, studies into the best performing solar materials have revealed a number of key properties likely to result in efficient devices [9]. Crucially, many of these properties, while difficult to measure experimentally, can be obtained relatively cheaply from theoretical methods, highlighting the importance of a combined experimental/theoretical approach when screening novel materials.

1.3.1   Magnitude and Nature of the Band Gap

Perhaps the most important fundamental property of a solar absorber is its band gap. The Shockley-Queisser limit [10], which takes into account the antagonistic dependence of the short-circuit current (J_{sc}) and open-circuit voltage (V_{oc}) on the band gap and solar spectrum (Fig. 1.6), provides a theoretical cap on the maximum obtainable efficiency for a material. The highest performing absorbers possess direct band gaps in the range 1.0–1.6 eV and while other materials may still be of interest, a band gap as close to 1.3 eV is desired in order to maximise efficiency at around 33%.

Fig. 1.6 The optimal range of photovoltaic absorber band gaps (shaded) projected on to the AM 1.5 solar spectrum (yellow) and the Shockley–Queisser limit (blue)

Fig. 1.7 The importance of a direct fundamental band gap—crystalline silicon (c-Si, purple) has drastically reduced absorption compared to $CH_3NH_3pbl_3$ (MAPI, red) and GaAs (green)

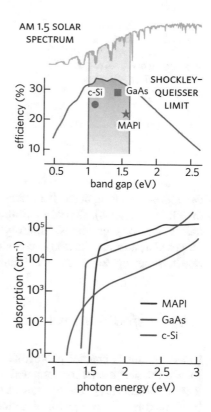

1.3.2   Strength of Optical Absorption

Strong optical absorption—characterised by a steep absorption edge in the absorption coefficient, α, $10 > ^4$–10^5 cm^{-1}—is crucial if a solar absorber is to achieve high efficiencies (Fig. 1.7) [11]. Indeed, many compounds with ideal band gaps perform poorly due to weak absorption, typically as a result of an indirect band gap, such as in crystalline silicon. However, as recently highlighted by Yu and Zunger, indirect band gap materials may still perform well if a direct transition with an optimal energy is also available [12]. Further reduction in absorption can occur if the fundamental band gap is dipole disallowed, thereby widening the optical band gap relative to the fundamental gap, as often seen in centrosymmetric systems [13, 14].

1.3.3   Charge-Carrier Effective Mass

Large charge-carrier mobilities enable efficient electron–hole separation and are essential for high-performance devices. The mobility of electrons and holes, μ, is related to the dispersion of the band edges in a material, with greater dispersion—or

Fig. 1.8 Band structure
schematic demonstrating
how increased band
dispersion (curvature) results
in smaller effective masses

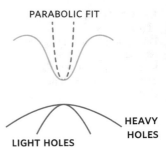

PARABOLIC FIT

HEAVY
HOLES

LIGHT HOLES

curvature—resulting in smaller effective masses and, in turn, greater charge-carrier mobilities (Fig. 1.8). Carrier mobilities are primarily limited through phonon and defect scattering processes. Minority-carrier lifetimes, τ, have also recently been proposed as a further transport screening metric for novel PV materials, due to their role in Shockley–Read–Hall recombination [9, 15, 16].

1.3.4 Defect Tolerance

Defect tolerance enables photovoltaics to retain strong optoelectronic properties, such as efficient carrier transport and power conversion efficiencies, despite the existence of defects [17]. The bonding structure shown in Fig. 1.9, whereby the valence band maximum is composed of antibonding interactions and the conduction band minimum is formed of bonding interactions, is thought to promote high levels of defect tolerance. In particular, it has been proposed that any dangling bonds formed due to vacancy defects will produce shallow states close to the band edges, rather than deep states in the middle of the band gap [18].

1.3.5 Dielectric Constant and Ferroelectric Behaviour

The electric response of a material is also vital in determining its photovoltaic performance. A large static dielectric constant indicates a high level of charge screening, enabling smaller defect charge-capture cross sections and reduced levels of non-radiative electron–hole recombination. Additionally, in the case of "hydrogenic" defects, a large dielectric constant enables smaller defect binding energies, resulting in shallower defect levels [19]. Ferroelectric photovoltaics have undergone a revival in recent months due to the potential link between high levels of hysteresis and ferroelectricity in emerging hybrid halide perovskite absorbers [20–24].

Fig. 1.9 The importance of
bonding structure on defect
tolerance: an electronic
structure with antibonding
states at the valence band
maximum is thought to
promote shallow over deep
defects. Adapted from [18]

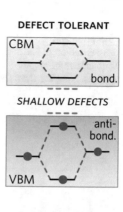

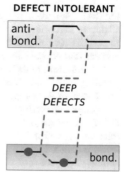

1.3.6 Rashba Splitting

Hybrid-halide perovskite absorbers have recently made their mark on the emergent
field of 'spintronics', due to the large Rashba and Dresselhaus splitting seen in
their electronic band structures [25]. Such a "spin-split indirect band gap" has been
shown, both theoretically and experimentally, to dramatically reduce the rate of
radiative recombination by a factor of more than 350% [26, 27]. Here, recombination
is suppressed by a mismatch in electron momentum between the conduction band
minima and valence band maximum, and is thought to enable the long charge-carrier
lifetimes seen in the hybrid perovskites (Fig. 1.10) [25, 28].

1.3.7 Band Alignment

When manufacturing solar devices, judicious choice of electron and hole contact
materials is essential to ensure favourable band alignment with the photovoltaic
absorber. By obtaining an efficient band alignment, the open circuit voltage can be
maximised and optimal carrier transport promoted (Fig. 1.11). In order to reduce

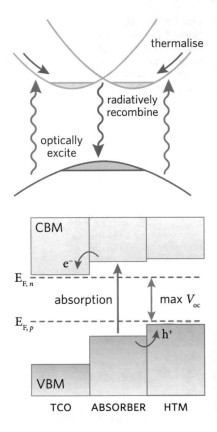

Fig. 1.10 The processes of absorption and recombination in Rashba and Dresselhaus spin-split systems. Adapted from [26]

Fig. 1.11 Schematic of band alignment in a heterojunction solar cell. Close alignment between components allows for maximisation of the open circuit voltage (V_{oc}). The Fermi level of the n-type transparent conducting oxide (TCO) and p-type hole transporting material (HTM) layers is denoted by $E_{F,n}$ and $E_{F,p}$, respectively

costs and streamline the manufacturing process, ideal solar absorbers should align well with ubiquitous contact materials, such as indium tin oxide and TiO_2.

1.4 Current Photovoltaic Absorbers

Most commercially available solar cells contain crystalline silicon as the absorber layer. Crystalline silicon devices have benefited from over four decades of research and possess efficiencies up to 26% [29], however, their performance is limited by the indirect band gap of silicon, resulting in low absorption coefficients [30]. Despite advanced manufacturing techniques such as localised back contacts and textured surfaces designed to maximise light capture, the efficiency of silicon devices has seen little improvement over the last five years. In addition, while the cost-per-panel of silicon cells has fallen steadily in recent years, this is mostly due to optimised fabrication methods and improved economies of scale [31, 32]. Further reduction in the price per Watt of panels is constrained by the underlying cost of producing silicon wafers, which is unlikely to change after over half a century of development in the transistor

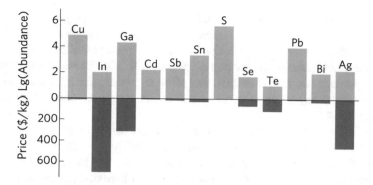

Fig. 1.12 Comparison of elemental cost and abundance in the earth's crust. Data from the London Metal Exchange

industry. These challenges have lead to the realisation of *thin-film* photovoltaics, in which use of a strongly absorbing material allows for thinner devices thereby reducing fabrication costs. These have shown record efficiencies up to 28.8%—for GaAs devices—but, due to the extremely high price of raw materials (Fig. 1.12), are limited to extraterrestrial applications where efficiency per gram takes precedence over cost [33, 34].

1.4.1   Third Generation Solar Absorbers

The drawbacks of commercially available photovoltaics have prompted research into possible alternatives, collectively termed *third generation* materials. In order to be competitive with crystalline silicon, these materials require comparable efficiencies combined with dramatically reduced manufacturing costs. This has been approached primarily through use of low-cost and earth-abundant materials coupled with solution-processing synthesis methods. Third-generation devices include those based on Cu_2ZnSnS_4 (CZTS) and dye-sensitised solar cells employing a liquid electrolyte. Unfortunately, despite over ten years of development both technologies suffer from poor efficiencies that limit their commercial potential.

1.4.2   Hybrid Halide Perovskites

In 2009, lead hybrid halide perovskites emerged as a novel class of solar absorber, sparking a global race to ever-more-efficient devices [35, 36]. Cells containing the champion hybrid halide perovskite, $CH_3NH_3pbl_3$ (MAPI), have been the subject of huge research interest, reaching efficiencies up to 22.7% in 2018 [37]. These record efficiencies have been enabled by a unique combination of properties, including a direct band gap of 1.55 eV [38], small exciton binding energies [39], high levels of defect self-regulation [40], remarkably long charge-carrier diffusion lengths [41, 42], and excellent charge-carrier mobilities [43, 44]. MAPI based devices are also cheap to produce due to low-temperature solution-processable synthesis routes and favourable band alignment with many commonly used hole and electron contact materials [45–47].

Unfortunately, the commercialisation of hybrid halide solar cells is currently restricted by several key issues: critically, the sensitivity of the perovskite structure toward oxidation by water has led to the need for strict device encapsulation, necessarily increasing manufacturing costs [36]. Furthermore, a growing number of reports suggest that the MAPI structure is *intrinsically* unstable, leading to decomposition even in the absence of moisture [48]. As such, there are concerns over the leaching of lead into the environment posing a risk to human and aquatic life. While many studies have attempted to increase the stability of MAPI, this has so far proved challenging and remains a significant issue facing the hybrid perovskite community [49, 50].

1.4.3 Emerging Absorbers

The success of the hybrid perovskites has energised the photovoltaic community. That a hitherto unknown material can surpass all other third-generation materials in such short a period of time highlights the potential for an earth-abundant and low-cost alternative to crystalline silicon. This rapid rise in performance has been based on an interdisciplinary approach combining a decades worth of developments in the synthesis, engineering, and photonics communities. Nevertheless, MAPI displays a fortuitous set of properties that enable remarkably efficient devices from films with high defect concentrations. As such, the speed of MAPI's success is likely to be an exception rather than a precedent to expect from all emerging absorbers. Indeed, any novel photovoltaic will likely require significant development before it can match the efficiency of the hybrid perovskites.

The last few years have seen the development of a broad array of emerging photovoltaic materials, several of which are highlighted in Fig. 1.13. In particular, the layered hybrid perovskites, along with a variety of absorbers containing bismuth, antimony, and tin have shown promise, with significant efficiency improvements seen in the last 2–3 years. As such, there is currently tremendous scope for the discovery and optimisation of emerging solar absorbers.

1.5 Thesis Outline

This thesis investigates a number of materials for their suitability as solar absorbers for use in photovoltaics. Throughout, ab initio methods are used to provide insight into the structural, electronic, and optical properties that determine real-world perfor-

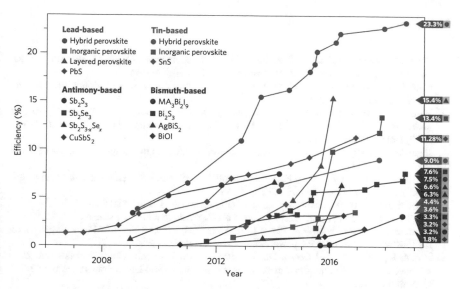

Fig. 1.13 Power conversion efficiency improvements over time for a selection of emerging photovoltaic absorbers. Lead-, tin-, antimony-, and bismuth-based absorbers are shown in orange, green, blue, and red, respectively

mance. Special attention is paid to understanding the behaviour of intrinsic defects due to their critical role in determining recombination and charge transport.

The rest of Part 1 examines the theoretical basis and methodology of the calculations undertaken. The quantum theory underpinning the approach is discussed, including the limitations and advantages of the various methods used, before the implementation of these methods is outlined. In particular, the processes for obtaining accurate predictions of important photovoltaic properties, such as the band gap and optical absorption, are presented.

Part 2 is concerned with perovskite-related absorbers. The intrinsic instability of MAPI is addressed, before several strategies for improving the stability of perovskite-based devices are proposed. A novel layered perovskite is examined, based on which the existence of a new family of layered-perovskite materials—expected to show high efficiencies with improved thermodynamic stability—is predicted. In addition, the vacancy-ordered double perovskites are examined as potential lead-free perovskite photovoltaics.

In Part 3, the search for emerging photovoltaic absorbers is expanded beyond the hybrid perovskite family. The bismuth chalcohalides are investigated due to their earth-abundant and non-toxic nature. The electronic and defect properties of BiSI and BiSeI are calculated and used to map the rate of Shockley-Read-Hall recombination across chemical potential space. The implications of these results for experimental groups aiming to produce efficient bismuth chalcohalide photovoltaic devices is addressed. The final chapter of this thesis contains a summary of all findings and discusses the impact of this work on the field of photovoltaics.

References

1. Khan MA, Khan MZ, Zaman K, Naz L (2014) Global estimates of energy consumption and greenhouse gas emissions. Renew Sustain Energy Rev 29:336–344
2. Martens P (2014) Health and climate change: modelling the impacts of global warming and ozone depletion. Routledge
3. World Health Organization, 7 million premature deaths annually linked to air pollution (2014). http://www.who.int/phe/health_topics/outdoorair/databases/FINAL_HAP_AAP_BoD_24March2014.pdf
4. Jäger-Waldau A (2007) Photovoltaics and renewable energies in Europe. Renew Sust Energy Rev 11:1414–1437
5. Lewis NS, Nocera DG (2006) Powering the planet: chemical challenges in solar energy utilization. Proc Natl Acad Sci 103:15729–15735
6. Hermann WA (2006) Quantifying global exergy resources. Energy 31:1685–1702
7. Nelson J (2003) The physics of solar cells, vol 57. World Scientific, Singapore
8. Ondraczek J, Komendantova N, Patt A (2015) WACC the dog: the effect of financing costs on the levelized cost of solar PV power. Renew Energy 75:888–898
9. Brandt RE, Stevanović V, Ginley DS, Buonassisi T (2015) Identifying defect-tolerant semiconductors with high minority-carrier lifetimes: beyond hybrid lead halide perovskites. MRS Commun 5:1–11
10. Shockley W, Queisser HJ (1961) Detailed balance limit of efficiency of p-n junction solar cells. J Appl Phys 32:510–519
11. De Wolf S, Holovsky J, Moon S-J, Löper P, Niesen B, Ledinsky M, Haug F-J, Yum J-H, Ballif C (2014) Organometallic halide perovskites: sharp optical absorption edge and its relation to photovoltaic performance. J Phys Chem Lett 5:1035–1039
12. Yu L, Zunger A (2012) Identification of potential photovoltaic absorbers based on first-principles spectroscopic screening of materials. Phys Rev Lett 108:068701
13. Kehoe AB, Scanlon DO, Watson GW (2011) Nature of the band gap of Tl_2O_3. Phys Rev B 83:3–6
14. Ganose AM, Scanlon DO (2016) Band gap and work function tailoring of SnO_2 for improved transparent conducting ability in photovoltaics. J Mater Chem C 4:1467–1475
15. Mattheis J, Werner JH, Rau U (2008) Finite mobility effects on the radiative efficiency limit of p-n junction solar cells. Phys Rev B 77:085203
16. Jaramillo R, Sher M-J, Ofori-Okai BK, Steinmann V, Yang C, Hartman K, Nelson KA, Lindenberg AM, Gordon RG, Buonassisi T (2016) Transient terahertz photoconductivity measurements of minority-carrier lifetime in tin sulfide thin films: advanced metrology for an early stage photovoltaic material. J Appl Phys 119:035101
17. Maughan AE, Ganose AM, Bordelon MM, Miller EM, Scanlon DO, Neilson JR (2016) Defect tolerance to intolerance in the vacancy-ordered double perovskite semiconductors Cs_2SnI_6 and Cs_2TeI_6. J Am Chem Soc 138:8453–8464
18. Zakutayev A, Caskey CM, Fioretti AN, Ginley DS, Vidal J, Stevanovic V, Tea E, Lany S (2014) Defect tolerant semiconductors for solar energy conversion. J Phys Chem Lett 5:1117–1125
19. Bube RH (1992) Photoelectronic properties of semiconductors. Cambridge University Press, Cambridge
20. Kutes Y, Ye L, Zhou Y, Pang S, Huey BD, Padtures NP (2014) Direct observation of ferroelectric domains in solution-processed $CH_3NH_3PbI_3$ perovskite thin films. J Phys Chem Lett 5:3335–3339
21. Liu S, Zheng F, Koocher NZ, Takenaka H, Wang F, Rappe AM (2015) Ferroelectric domain wall induced band gap reduction and charge separation in organometal halide perovskites. J Phys Chem Lett 6:693–699
22. Frost JM, Butler KT, Brivio F, Hendon CH, Van Schilfgaarde M, Walsh A (2014) Atomistic origins of high-performance in hybrid halide perovskite solar cells. Nano Lett 14:2584–2590
23. Butler KT, Frost JM, Walsh A (2015) Band alignment of the hybrid halide perovskites $CH_3NH_3PbCl_3$, $CH_3NH_3PbBr_3$ and $CH_3NH_3PbI_3$. Mater Horiz 2:228–231

24. Shao Y, Xiao Z, Bi C, Yuan Y, Huang J (2014) Origin and elimination of photocurrent hysteresis by fullerene passivation in $CH_3NH_3PbI_3$ planar heterojunction solar cells. Nat Commun 5:5784
25. Kepenekian M, Robles R, Katan C, Sapori D, Pedesseau L, Even J (2015) Rashba and dresselhaus effects in hybrid organic-inorganic perovskites: from basics to devices. ACS Nano 9:11557–11567
26. Azarhoosh P, Frost JM, McKechnie S, Walsh A, van Schilfgaarde M (2016) Relativistic origin of slow electron-hole recombination in hybrid halide Perovskite solar cells. APL Mater 4:091501
27. Hutter EM, Gélvez-Rueda MC, Osherov A, Bulović V, Grozema FC, Stranks SD, Savenije TJ (2016) Direct-indirect character of the bandgap in methylammonium lead iodide perovskite. Nat Mater 2016 (In press)
28. Etienne T, Mosconi E, Angelis FD (2016) Dynamical origin of the rashba effect in organohalide lead perovskites: a key to suppressed carrier recombination in Perovskite solar cells? J.Phys Chem Lett 7:1638–1645
29. Yoshikawa K, Kawasaki H, Yoshida W, Irie T, Konishi K, Nakano K, Uto T, Adachi D, Kanematsu M, Uzu H, Yamamoto K (2017) Silicon heterojunction solar cell with interdigitated back contacts for a photoconversion efficiency over 26%. Nat Energy 2:17032
30. Peng J, Lu L, Yang H (2013) Review on life cycle assessment of energy payback and greenhouse gas emission of solar photovoltaic systems. Renew Sust Energy Rev 19:255–274
31. Lacerda JS, van den Bergh JC (2016) Diversity in solar photovoltaic energy: implications for innovation and policy. Renew Sustain Energy Rev 54:331–340
32. Rubin ES, Azevedo IM, Jaramillo P, Yeh S (2015) A review of learning rates for electricity supply technologies. Energy Policy 86:198–218
33. Wadia C, Alivisatos AP, Kammen DM (2009) Materials availability expands the opportunity for large-scale photovoltaics deployment. Environ Sci Technol 43:2072–2077
34. Dhere NG (2007) Toward GW/year of CIGS production within the next decade. Sol Energy Mater Sol Cells 91:1376–1382
35. Green MA, Ho-Baillie A, Snaith HJ (2014) The emergence of perovskite solar cells. Nat Photon 8:506–514
36. Grätzel M (2014) The light and shade of perovskite solar cells. Nat Mater 13:838–842
37. Green M, Yoshihiro H, Dunlop ED, Levi DH, Hohl-Ebinger J, Ho-Baillie AWY (2017) Solar cell efficiency tables (version 51). Prog Photovoltaics Res Appl 26:3–12
38. Baikie T, Fang Y, Kadro JM, Schreyer M, Wei F, Mhaisalkar SG, Graetzel M, White TJ (2013) Synthesis and crystal chemistry of the hybrid perovskite $(CH_3NH_3)PbI_3$ for solid-state sensitised solar cell applications. J Mater Chem A 1:5628–5641
39. D'Innocenzo V, Grancini G, Alcocer MJ, Kandada ARS, Stranks SD, Lee MM, Lanzani G, Snaith HJ, Petrozza A (2014) Excitons versus free charges in organo-lead tri-halide perovskites. Nat Commun 5:3586
40. Walsh A, Scanlon DO, Chen S, Gong X, Wei S-H (2015) Self-regulation mechanism for charged point defects in hybrid halide perovskites. Angew Chem 127:1811–1814
41. Deng Y, Peng E, Shao Y, Xiao Z, Dong Q, Huang J (2015) Scalable fabrication of efficient organolead trihalide perovskite solar cells with doctor-bladed active layers. Energy Environ Sci 8:1544–1550
42. Xing G, Mathews N, Sun S, Lim SS, Lam YM, Gratzel M, Mhaisalkar S, Sum TC (2013) Long-range balanced electron- and hole-transport lengths in organic-inorganic $CH_3NH_3PbI_3$. Science 342:344–347
43. Wehrenfennig C, Eperon GE, Johnston MB, Snaith HJ, Herz LM (2014) High charge carrier mobilities and lifetimes in organolead trihalide perovskites. Adv Mater 26:1584–1589
44. Zhao Y, Zhu K (2013) Charge transport and recombination in perovskite $(CH_3NH_3)PbI_3$ sensitized TiO_2 solar cells. J Phys Chem Lett 4:2880–2884
45. Barrows AT, Pearson AJ, Kwak CK, Dunbar ADF, Buckley AR, Lidzey DG (2014) Efficient planar heterojunction mixed-halide perovskite solar cells deposited via spray-deposition. Energy Environ Sci 7:2944
46. Bhachu DS, Scanlon DO, Saban EJ, Bronstein H, Parkin IP, Carmalt CJ, Palgrave RG (2015) Scalable route to $CH_3NH_3PbI_3$ perovskite thin films by aerosol assisted chemical vapour deposition. J Mater Chem A 3:9071–9073

47. Hodes G, Cahen D (2014) Photovoltaics perovskite cells roll forward. Nat Photon 8:87–88
48. Christians JA, Manser JS, Kamat PV (2015) Best practices in Perovskite solar cell efficiency measurements. Avoiding the error of making bad cells look good. J Phys Chem Lett 6:852–857
49. Guarnera S, Abate A, Zhang W, Foster JM, Richardson G, Petrozza A, Snaith HJ (2015) Improving the long-term stability of perovskite solar cells with a porous Al_2O_3 buffer layer. J Phys Chem Lett 6:432–437
50. Niu G, Guo X, Wang L (2015) Review of recent progress in chemical stability of Perovskite solar cells. J Mater Chem A 3:8970–8980

Chapter 2
Computational Theory

2.1 Quantum Mechanical Approaches

All calculations presented in this thesis were performed using ab initio methods.
These describe chemical systems by working up from fundamental physics, in contrast to empirical or semi-empirical approaches which rely on fitting to experimental measurements. The principle goal is finding a solution to the time-independent
Schrödinger equation [1–3],

$$\hat{H}\Psi = E\Psi, \tag{2.1}$$

where $\hat{H}$ is the Hamiltonian operator composed of kinetic and potential energy terms,
E is the system energy, and

$$\Psi = \Psi(\mathbf{r}_1, \ldots, \mathbf{r}_N, \mathbf{r}_1^n, \ldots, \mathbf{r}_M^n)$$

is a wavefunction that depends on the position of N electrons $\{\mathbf{r}\}$ and M nuclei $\{\mathbf{r}^n\}$.
The Hamiltonian is of the form

$$
\hat{H} = -\overbrace{\frac{\hbar^2}{2m_e}\sum_i \nabla_i^2}^{\hat{T}_{\text{elec}}} - \overbrace{\frac{\hbar^2}{2}\sum_k \frac{\nabla_k^2}{M_k}}^{\hat{T}_{\text{nuc}}} + \overbrace{\frac{1}{2}\sum_{i\neq j} \frac{e^2}{|\mathbf{r}_i - \mathbf{r}_j|}}^{V_{\text{elec-elec}}}
$$
$$
+ \underbrace{\frac{1}{2}\sum_{k\neq l} \frac{Z_k Z_l}{\mathbf{r}_k^n - \mathbf{r}_l^n}}_{V_{\text{nuc-nuc}}} - \underbrace{\sum_{i\neq k} \frac{e Z_k}{\mathbf{r}_i - \mathbf{r}_k^n}}_{V_{\text{nuc-elec}}}, \tag{2.2}
$$

where $\hbar$ is the reduced Planck constant, i is the index over all electrons, k is the
index over all nuclei, ∇^2 is the Laplacian operator, e is the electron charge, m_e is the
electron rest mass, M is the nuclear mass, and Z is the nuclear charge.

© The Editor(s) (if applicable) and The Author(s), under exclusive license
to Springer Nature Switzerland AG 2020
A. Ganose, *Atomic-Scale Insights into Emergent Photovoltaic Absorbers*,
Springer Theses, https://doi.org/10.1007/978-3-030-55708-9_2

Theoretically, finding an exact solution to the Schrödinger equation would allow one to calculate the exact ground state energy and wavefunction of a system. However, an analytic solution to the Schrö-dinger equation is only possible for systems containing a single atom with a single electron. While numerical methods allow for very accurate solutions up to several electrons, beyond this the number of parameters that must be optimised is so large as to be computationally intractable.[1] As such, several approximations must be made, with the hope that the model retains physical relevance.

The first approximation is that the nuclei are considered stationary with respect to the movement of the electrons. This is termed the *Born–Oppenheimer approximation* and is generally valid given that the mass of the proton is ~2000 times the mass of the electron. This allows us to ignore the kinetic energy of the nuclei and rewrite the Hamiltonian as

$$\hat{H} = \hat{T}_{\text{elec}} + V_{\text{nuc}-\text{elec}} + V_{\text{elec}-\text{elec}} + V_{\text{nuc}-\text{nuc}}, \qquad (2.3)$$

where the potential energy of the nuclear interactions, $V_{\text{nuc}-\text{nuc}}$, is simply a constant for a static set of atomic coordinates [3]. The terms describing the kinetic energy of the electrons, $\hat{T}_{\text{elec}}$, and the nuclear–electron interactions, $V_{\text{nuc}-\text{elec}}$, are relatively simple to evaluate. Calculating the potential energy between electrons, $V_{\text{elec}-\text{elec}}$, however, presents a significant challenge due to the *many-body problem*.

The variational principle is crucial to obtaining numerical solutions to the Schrödinger equation. It states that the expectation value of a Hamiltonian calculated using a trial wavefunction, Ψ_T, will always be greater than the true ground state energy, E_0, produced from the true wavefunction, Ψ_0:

$$\langle \Psi_T | \hat{H} | \Psi_T \rangle = E_T \geq E_0 = \langle \Psi_0 | \hat{H} | \Psi_0 \rangle. \qquad (2.4)$$

Accordingly, by using an increasing number of trial wavefunctions we can obtain a more faithful description of the true wavefunction and convergence towards a more accurate total energy.

2.2 Hartree–Fock Method

To avoid the challenges posed by the many-body problem, in 1928 Hartree posited that the full multi-electron wavefunction could be approximated as a set of one-electron wavefunctions (χ) termed the Hartree product,

$$\Psi(\mathbf{x}_1, \mathbf{x}_2, \ldots, \mathbf{x}_N) = \chi_1(\mathbf{x}_1)\chi_2(\mathbf{x}_2) \ldots \chi_N(\mathbf{x}_N), \qquad (2.5)$$

---

[1] Walter Kohn, a pioneer of quantum mechanical modelling, termed this challenge an "exponential wall" in his 1999 Nobel Prize lecture [4]. Indeed, for a system containing 100 electrons, one would need to minimise a parameter space with -10^{150} dimensions.

where $\mathbf{x}_i$ represents a generalised coordinate including the x, y, and z cartesian coordinates and spin. He further introduced the idea of a one-electron Hamiltonian operator, $\hat{h}$, which acted on a one-electron wavefunction to produce the eigenequation

$$\hat{h}_i \chi_i(\mathbf{x}_i) = \varepsilon_i \chi_i(\mathbf{x}_i), \qquad (2.6)$$

where i is an electron occupying a single spin-orbital χ_i, with energy ε_i, and the one-electron Hamiltonian takes the form

$$\hat{h}_i = -\frac{\hbar^2}{2m_e} \sum_i \nabla_i^2 + \sum_k \frac{e Z_k}{|\mathbf{r}_i - \mathbf{r}_k^n|}. \qquad (2.7)$$

The total energy of this non-interacting system is defined as the sum of the one-electron energies.

Due to the omission of electron–electron interactions, this model fails to realistically describe the behaviour of most systems. Crucially, the Hartree product does not obey the Pauli exclusion principle—namely, that the wavefunction is antisymmetric with respect to the exchange of two fermions:

$$\Psi(\ldots, \mathbf{x}_1, \mathbf{x}_2, \ldots) = -\Psi(\ldots, \mathbf{x}_2, \mathbf{x}_1, \ldots). \qquad (2.8)$$

Subsequent work by Fock and Slater exploited the well-known sign-changing property of matrix determinants [3, 5]. Rather than constructing the electronic wavefunction as a product of individual spin-orbitals, they noted that a single Slater determinant will instead produce a wavefunction that obeys the Pauli principle. For example, in a two-electron system:

$$\Psi(\mathbf{x}_1, \mathbf{x}_2) = \frac{1}{\sqrt{2}} \begin{vmatrix} \chi_1(\mathbf{x}_1) & \chi_2(\mathbf{x}_1) \\ \chi_1(\mathbf{x}_2) & \chi_2(\mathbf{x}_2) \end{vmatrix} = \frac{1}{\sqrt{2}} [\chi_1(\mathbf{x}_1)\chi_2(\mathbf{x}_2) - \chi_1(\mathbf{x}_2)\chi_2(\mathbf{x}_1)] \qquad (2.9)$$

The antisymmetric nature of the wavefunction can be seen if the two electrons are exchanged:

$$\begin{aligned} \Psi(\mathbf{x}_2, \mathbf{x}_1) &= \frac{1}{\sqrt{2}} [\chi_1(\mathbf{x}_2)\chi_2(\mathbf{x}_1) - \chi_1(\mathbf{x}_1)\chi_2(\mathbf{x}_2)] \\ &= -\frac{1}{\sqrt{2}} [\chi_1(\mathbf{x}_1)\chi_2(\mathbf{x}_2) - \chi_1(\mathbf{x}_2)\chi_2(\mathbf{x}_1)] \\ &= -\Psi(\mathbf{x}_1, \mathbf{x}_2) \end{aligned} \qquad (2.10)$$

Based on this, Hartree and Fock introduced two further terms to allow for calculation of the Hartree–Fock energy:

$$J_{ij} = \frac{1}{2} \iint \chi_i(\mathbf{x}_1) \chi_j^*(\mathbf{x}_2) \frac{e^2}{|\mathbf{r}_1 - \mathbf{r}_2|} \chi_i(\mathbf{x}_1) \chi_j^*(\mathbf{x}_2) d\tau_1 d\tau_2 \qquad (2.11)$$

$$K_{ij} = \frac{1}{2} \iint \chi_i(\mathbf{x}_1) \chi_j^*(\mathbf{x}_2) \frac{e^2}{|\mathbf{r}_1 - \mathbf{r}_2|} \chi_j(\mathbf{x}_1) \chi_i^*(\mathbf{x}_2) d\tau_1 d\tau_2, \qquad (2.12)$$

where integrals with respect to $d\tau$ represent integration over all space, J_{ij} is the Coulomb term representing the electrostatic repulsion between electrons, and K_{ij} is the exchange integral representing a quantum mechanical effect arising from the antisymmetric wavefunction. Electron exchange is a purely quantum mechanical effect driven by the Pauli principle. The Slater determinant wavefunctions ensure the exchange term will only be non-zero for electrons of the same spin, with the resulting exchange energy manifesting as a negative correction to the Coulomb energy.

The Hartree-Fock equations can then be written

$$\left[\hat{h}_i + \sum_{i \neq j} J_{ij} - \sum_{i \neq j} K_{ij} \right] \chi_i = \varepsilon_i \chi_i. \qquad (2.13)$$

The formulation of the Coulomb term simplifies the many-body problem of interacting electrons by adopting the *mean field approximation*. Here, each electron will move in an average potential field generated by the presence of all other independent electrons. This allows one to solve the Hartree-Fock equations iteratively. Initially, a trial set of wavefunctions is chosen, from which the potential field for each electron is calculated. As the fields depend on the spin-orbitals of the other electrons, solving Eq. (2.13) will result in a new set of spin-orbitals that can in turn be used to obtain a new set of potential fields. This procedure is repeated until the fields no longer change, resulting in a *self-consistent field*. To solve the entire system, the variational principle is employed to find the set of wavefunctions that produce the lowest energy.

While the Hartree–Fock method is still used today, its adoption of the mean-field approach results in a number of limitations. In particular, as each electron moves independently of the other electrons in the system, electron correlation—the property of electrons whereby they "avoid" each other—is ignored [6]. As such, Hartree–Fock fails for systems containing localised electrons, such as transition metals. Furthermore, while the energy difference due to electron correlation generally amounts to only ~0.1% of the total energy of a system, this is similar in magnitude to binding and reaction energies and can therefore lead to incorrect predictions.

2.3   Density Functional Theory

Density functional theory (DFT) takes an alternative approach in attempting to solve the total energy of a system. In 1964, Hohenberg and Kohn demonstrated that the ground state energy, along with all other ground state properties, is uniquely defined by the electron density, $\rho(\mathbf{r})$ [7]. In contrast to Hartree–Fock theory, which relies on solving the wavefunction, density functional theory attempts to solve for the electron density. This presents a significant simplification as the density depends on just 3 variables (x, y, and z coordinates), whereas the wavefunction depends on $3N$ variables, where N is the total number of electrons in the system. The variational principle again applies, as the total energy calculated from the trial density will always be greater or equal to the true ground state energy.

Hohenberg and Kohn proved that the total energy functional can be written

$$E[\rho(\mathbf{r})] = \int \rho(\mathbf{r}) V_{ext}(\mathbf{r}) d\mathbf{r} + F_{HK}[\rho(\mathbf{r})], \qquad (2.14)$$

where V_{ext} is the external potential created by the nuclei and any external fields, and F_{HK} is the Hohenberg and Kohn *universal functional* containing all information about the kinetic energies and electron–electron interactions. Should the form of the universal functional be known, theoretically it should be possible to access the exact ground state properties of any system. In 1965, Kohn and Sham suggested a scheme for approximating F_{HK} by considering a fictitious non-interacting system of quasiparticles with an identical density to that of the fully interacting system of electrons. They started by separating the universal functional into interacting and non-interacting terms

$$F_{HK}[\rho(\mathbf{r})] = E_H[\rho(\mathbf{r})] + T_S[\rho(\mathbf{r})] + E_{xc}[\rho(\mathbf{r})], \qquad (2.15)$$

where E_H is the classical (non-interacting) Hartree energy calculated as

$$E_H = \frac{e^2}{2} \iint \frac{\rho(\mathbf{r}_i)\rho(\mathbf{r}_j)}{|\mathbf{r}_i - \mathbf{r}_j|} d\mathbf{r}_i d\mathbf{r}_j, \qquad (2.16)$$

T_S is the kinetic energy of the non-interacting electrons, and E_{xc} is the sum of the non-classical electron–electron interactions and a correction to the kinetic energy accounting for non-classical interactions (thereby yielding the correct total kinetic energy).

To represent this system of non-interacting quasiparticles, Kohn and Sham reintroduced the idea of a wavefunction comprising a Slater determinant of *Kohn–Sham wavefunctions*, ψ_i. Crucially, this allowed for the exact solution of the non-interacting kinetic energy in the same manner as in Hartree–Fock

$$T_S[\rho(\mathbf{r})] = -\frac{\hbar^2}{2m_e} \sum_i \langle \psi_i | \nabla^2 | \psi_i \rangle. \qquad (2.17)$$

The density could therefore be written as the sum of the densities of the Kohn–Sham wavefunctions, which exactly recreate the density from the true multibody wavefunction

$$\rho(\mathbf{r}) = \sum_i |\psi_i(\mathbf{r})|^2 = |\Psi_{\mathrm{mb}}(\mathbf{r})|^2. \qquad (2.18)$$

By applying the appropriate variational condition, the one-electron Kohn–Sham equations can be written

$$\left[-\frac{\hbar^2}{2m_e}\nabla^2 + e^2 \int \frac{\rho(\mathbf{r}')}{|\mathbf{r}-\mathbf{r}'|}\mathrm{d}\mathbf{r} - \sum_k \frac{eZ_k}{|\mathbf{r}-\mathbf{r}_k^n|} + V_{\mathrm{xc}}[\rho(\mathbf{r})] \right] \psi_i(\mathbf{r}) = \varepsilon_i \psi_i(\mathbf{r}),$$

$$(2.19)$$

where V_{xc} is the exchange–correlation functional related to the exchange-correlation energy, E_{xc}.[2] The ground state energy may then be solved in a self-consistent manner similar to as in the Hartree–Fock method. Initially, a set of trial wavefunctions are generated from which the density can be determined. This density is fed into Eq. (2.19) to produce a new set of Kohn–Sham wavefunctions and the process repeated until convergence is achieved.

While density functional theory is exact if the exact exchange–correlation functional is known, the true functional form has not been elucidated and therefore must be approximated. As such, unlike Hartree–Fock which contains exact exchange and zero correlation, DFT instead obtains approximate exchange and approximate correlation. Despite this, DFT is widely employed in computational semiconductor research in both physics and chemistry communities.

2.4 Density Functional Theory: Implementation

2.4.1 Exchange–Correlation Functionals

The design of the exchange–correlation functional is the primary factor in determining the accuracy of a DFT calculation. One of DFT's successes is that simple approximations to $E_{\mathrm{xc}}[\rho(\mathbf{r})]$ can produce accurate results. The most basic exchange-correlation functionals employ the *local-density approximation* (LDA), in which the amount of exchange and correlation at a given point is based solely on the *local* density—namely, the density of electrons at that point. The values of exchange and correlation have been extracted from high-precision quantum-Monte-Carlo simulations of uniform electron gases—hypothetic systems in which a number of electrons are free to move in an electrically neutral environment [8–10]. In particular, the parameterisation of these calculations by Vosko et al. [11] is often employed by DFT packages. The LDA is best suited for systems in which the electron density varies slowly but suffers from a number of failures when elements with tightly bound valence electrons, such as transition metals, are introduced. These manifest

---

[2]It should be noted that the *Kohn–Sham orbitals* obtained from Eq. (2.19) need not resemble traditional chemical orbitals. Indeed, they can be *any* set of functions, provided they result in a density that minimises the total energy.

as an overbinding of electrons resulting in the overestimation of atomisation and ionisation energies and the underestimation of bond lengths.

Gradient-corrected functionals attempt to improve on the LDA by including not only the local density, $\rho(\mathbf{r})$, but also the gradient of the density, $\nabla\rho(\mathbf{r})$. A large variety of formulations exist; some are based solely on first-principles approaches whereas others are fitted to experimental data (*semi-empirical* functionals). In general, those which rely on local gradients within cutoff regions are termed *generalised-gradient approximation* (GGA) functionals [12].

2.4.1.1 PBE and PBEsol

Perdew, Burke, and Ernzerhof introduced their popular GGA functional, PBE, in 1996 [13]. PBE provides several improvements for periodic systems when compared to other GGA functionals but slightly underbinds electrons and therefore overestimates bond lengths [14]. The PBEsol functional was released by Perdew and coworkers in 2008 as a version of the PBE functional revised for solids [15]. By providing the functional with more "LDA-like" behaviour, PBEsol occupies a happy medium between the LDA and PBE, producing both better energies and lattice parameters in periodic systems [14]. PBEsol is employed extensively throughout this thesis.

Despite their widespread adoption, the LDA and GGA suffer from several issues which limit their accuracy. Primarily, both display the *self-interaction error* (SIE) in which an electron may spuriously interact with itself. This arises due to the use of the mean field approximation coupled with an approximate exchange–correlation functional. In contrast, Hartree–Fock avoids the SIE due to the exchange integral and Slater determinant basis (i.e. the presence of exact exchange) which prevent two electrons with the same spin from occupying the same position. This issue is most apparent for systems containing highly localised orbitals, such the first-row transition metals where there are a greater number of electrons in close proximity. The SIE leads to unphysical delocalisation of electrons in an attempt to minimise the degree of self-interaction. The use of approximate exchange and correlation also results in the underestimation of semiconductor band gaps—occasionally predicting known semiconductors to be metallic [16, 17].

2.4.1.2 Hybrid DFT

Hybrid functionals attempt to correct for the inaccuracies inherent in the LDA and GGA by introducing an amount of exact exchange from Hartree–Fock. Due to the expense of calculating the exchange term, this approach has only recently become computationally feasible in large periodic systems. Regardless, its use is quickly becoming the standard in the solid-state chemistry and physics communities. A commonly used non-parameterised hybrid functional is PBE0 [18], formulated as

$$E_{xc}^{PBE0} = E_{xc}^{PBE} + \alpha \left(E_x^{HF} - E_x^{PBE} \right),\qquad(2.20)$$

where α is 0.25 based on perturbation theory [19].

Due to the slow convergence of the exchange energy over distance, a number of range separated functionals have been developed in an attempt to reduce the computational cost associated with hybrid calculations. The Heyd–Scuseria–Ernzerhof (HSE06) functional is one such approach that has gained popularity [20, 21]. Here, only the short-range exchange energy is calculated using Hartree–Fock, with the remainder provided by PBE, as

$$E_{xc}^{HSE} = \alpha E_x^{HF,SR}(\omega) + (1-\alpha)E_x^{PBE,SR}(\omega) + E_x^{PBE,LR}(\omega) + E_c^{PBE},\qquad(2.21)$$

where ω is $0.11 a_0^{-1}$ and α is 0.25. Accordingly, the functional tends to PBE0 as $\omega \to \infty$ and PBE as $\omega \to 0$. While ultimately a form of error cancellation, hybrid functionals have been shown to provide significant improvements over purely local based approaches. In particular, both PBE0 and HSE06 are able to accurately reproduce the low-temperature experimental band gaps of many semiconductors, often to within 0.1 eV [22]. As HSE06 generally performs better for small and medium band gap materials such as photovoltaics, it has been employed for all high accuracy electronic structure calculations in this thesis.

2.4.2 Dispersion Corrections

Most density functionals are unable to account for dispersive van der Waals type interactions as these arise from dynamical correlations between fluctuating charge distributions. These are particularly important for low-dimensionality materials, such as those containing planar sheets or nano-ribbons. In the density functional dispersion correction formalism (DFT-D), an additional term is added to the conventional Kohn–Sham total energy [23], as

$$E_{DFT-D} = E_{KS} + E_{disp}.\qquad(2.22)$$

A large number of correction schemes have been proposed [24–28]. In this thesis we have employed the DFT-D3 approach suggested by [26] in 2010. Here, the dispersion correction is given by a pairwise sum over all atoms in the system (labelled by i and j), as

$$E_{disp} = -\frac{1}{2}\sum_{ij}\sum_{n=6,8}\frac{C_n^{ij}}{|\mathbf{r}_i - \mathbf{r}_j|^n}f_{d,n}(|\mathbf{r}_i - \mathbf{r}_j|),\qquad(2.23)$$

where C_n^{ij} are the averaged geometry-dependent nth-order dispersion coefficients for the ij atom pair, calculated based on fitting to time-dependent DFT calculations, and $f_{d,n}$ is a damping function to avoid singularities for small internuclear distances and double-counting at intermediate distance s, where the damping is parameterised based on the choice of exchange–correlation functional. In particular, we have adopted the zero damping D3 correction. DFT-D3 has been shown to result in improved forces and geometries for a wide range of solid state materials [26]. The Axilrod–Teller triple dipole terms were not included in this work.

2.4.3 Periodic Boundary Conditions

It is not practical to solve the Schrödinger equation for macroscopic materials due to the large number of particles they contain. We can instead take advantage of the *crystalline* nature of many solid-state materials by considering a much smaller repeating unit. Here the geometric arrangement of atoms in a macroscopic solid is approximated as an infinitely repeating perfect crystal. The smallest repeating set of atoms, termed the *motif*, is located on the lattice points of a three-dimensional *Bravais lattice*, which is periodically repeated in all three dimensions to form the overall structure (Fig. 2.1). The structure can be defined in terms of a *unit cell*, a parallelepiped cell that can be tiled to build the whole crystal. While there are an infinite number of possible unit cells for a given lattice, crystallographic conventions describe two classes of cell: the *primitive cell* is a unit cell containing exactly one lattice point, and the *conventional cell* is the smallest possible unit cell possessing the same symmetry as the overall lattice and may contain more than one lattice point. Both cells will produce the same crystal structure when repeated throughout space. To reduce computational cost the primitive cell is used for all calculations.

The cell is described by a set of lattice vectors, **a**, **b**, and **c**, with the lattice containing all periodic images given by

$$\mathbf{R} = n_1\mathbf{a} + n_2\mathbf{b} + n_3\mathbf{c}, \tag{2.24}$$

where $\{n_1, n_2, n_3\}$ are integers. The lattice vectors determine the cell shape, such as the cell angles α, β, and γ. While translational symmetry is required in all Bravais lattices, the cell may possess additional symmetry in the arrangement of the atoms. Accordingly, the wavefunction will experience the same symmetry constraints, often enabling a further reduction in computational cost.

The use of Bloch's theorem enables us to simplify the solution of the one-electron wavefunctions under periodic boundary conditions. It states that the wavefunction of particle in a periodic potential is formed of

(a) a function, $u_\mathbf{k}(\mathbf{r})$, with the periodicity of the lattice, and
(b) a plane wave $e^{i\mathbf{kr}}$.

The wavefunction of electron n with wave vector $\mathbf{k}$ may therefore be written

$$\psi_{n,\mathbf{k}}(\mathbf{r}) = u_{n,\mathbf{k}}(\mathbf{r})e^{i\mathbf{k}\cdot\mathbf{r}}. \tag{2.25}$$

Fig. 2.1 Representation of
the primitive cell (green),
conventional cell (orange),
and lattice (blue) of the face
centred cubic structure,
indicating the effects of
periodic boundary conditions

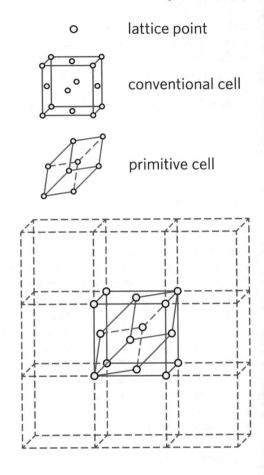

Due to the periodicity of the lattice, any wave vector $\mathbf{k}'$ that differs from $\mathbf{k}$ by
an integer number of reciprocal lattice vectors will result in an identical solution.
Accordingly we can restrict the values of $\mathbf{k}$ to those occurring in smallest repeating
unit of the reciprocal lattice, termed the first *Brillouin zone*.

When studying periodic systems through HF or DFT, one must therefore find the
set of periodic functions that minimises the total energy in a normalised sum over all
$\mathbf{k}$-vectors. As an infinite number of $\mathbf{k}$-vectors exist, an approximation must be made
in the form of discrete $\mathbf{k}$-point sampling. As the wavefunction varies smoothly within
$\mathbf{k}$-space, this approximation will hold provided the space is sampled with sufficient
density. In this thesis we have adopted the Monkhorst-Pack method, whereby an
equally spaced grid of points are sampled. In all cases the grid is centred at the origin
of reciprocal space, termed the Γ point. Semiconductors and insulators generally
require relatively low sampling densities, whereas metallic systems necessitate dense
$\mathbf{k}$-point meshes so that the Fermi level can be accurately determined. To ensure
calculations remain accurate without excess computational cost, the $\mathbf{k}$-point mesh
density is systematically increased until convergence criteria are met.

2.4.4 Basis Set and Pseudopotentials

The remaining challenge is to determine the form of the periodic potential, $u(\mathbf{r})$. In most ab initio approaches, the potential is represented as a linear combination of basis functions. The choice of these functions is a primary factor in determining the accuracy and cost of a calculation. A plane wave basis set is a natural choice for periodic calculations, in part due to their equivalent representation as a Fourier series [29]. As the plane wave basis is only dependent on the size of the simulation cell, it will not suffer from basis set superposition errors in contrast to localised basis sets such as Gaussian functions.

Since the potential has the same periodicity as the lattice, it can be expressed as a linear combination of plane waves with wavevectors $\mathbf{G}$ that are reciprocal lattice vectors of the crystal

$$u_n(\mathbf{r}) = \sum_{\mathbf{G}} c_{n,\mathbf{G}} e^{i\mathbf{G}\cdot\mathbf{r}}, \tag{2.26}$$

where $c_{n,\mathbf{G}}$ are the plane wave coefficients. Together with Eq. (2.25), the one-electron wavefunctions can be written

$$\psi_{n,\mathbf{k}}(\mathbf{r}) = \sum_{\mathbf{G}} c_{n,\mathbf{k}+\mathbf{G}} e^{i(\mathbf{k}+\mathbf{G})\cdot\mathbf{r}}. \tag{2.27}$$

While in principle use of a plane wave basis set is exact given an infinite number of plane waves, in practice a finite-sized basis set must be used. This is achieved by defining a plane wave energy cutoff, with the plane wave energy defined by

$$E_{\mathbf{k}+\mathbf{G}} = \frac{\hbar^2}{2m_e} |\mathbf{k} + \mathbf{G}|^2. \tag{2.28}$$

Large cutoff energies are needed to represent wavefunctions where the electron density oscillates rapidly—for example, the electrons near atomic cores. The cutoff leads to another advantage of using a plane wave basis set, in that the accuracy of the basis can controlled through a single parameter—a significant simplification compared to localised basis sets. Similar to the $\mathbf{k}$-space sampling density, the cutoff energy is increased until convergence is reached.

The wavefunctions of valence electrons oscillate rapidly close to the nucleus due to the strong nuclear potential and the requirement that they be orthogonal to the core electrons. Accordingly, a large number of high energy plane waves are required to accurately model these electrons, significantly increasing the computational cost. As the core electrons only interact weakly with the valence electrons, it is convenient to approximate their behaviour by use of an effective potential representing a screened nuclear core. This is achieved through use of *pseudopotentials*. These are smoothly varying functions that produce pseudo-wavefunctions with the same shape as the "true" wavefunction outside a certain cutoff but with fewer nodes within it (Fig. 2.2). The pseudopotentials are carefully fitted for each element based on all-

Fig. 2.2 Schematic
representation of the
relationship between an all
electron (blue dashed line)
and pseudopotential (solid
red line) wavefunction (Ψ)
and potential (V). The
psuedo-wavefunction
replicates the all-electron
wavefunction outside the
cutoff, r_{cut}

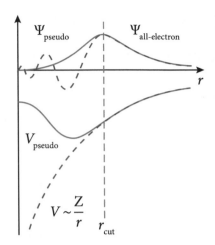

electron calculations, and are constructed such that the scattering properties of the
pseudo-wavefunctions are the same as the scattering produced by the ion and core
electrons. As such, pseudopotentials are generally angular momentum dependent.
Use of pseudopotentials makes calculations involving large atoms with many core
electrons tractable but reduces the accuracy of results [30].

The calculations performed in this thesis employ the projector augmented-wave
method (PAW). Developed by Blöchl in 1994 [31], this method aims to enable the
physics of all-electron calculations at a pseudopotential-like computational cost [32].
Here, the core electrons are represented by localised basis sets, with the effects of the
core states projected onto the valence electrons. As such, the correct nodal behaviour
of the valence electron wavefunctions is maintained. All-electron scalar relativistic
calculations are used to generate the PAW fixed core states.

References

1. Schrödinger E (1926) An undulatory theory of the mechanics of atoms and molecules. Phys
   Rev 28:1049
2. Schrödinger E (1926) Quantisierung als eigenwertproblem. Annalen der physik 385:437–490
3. Fock V (1930) Näherungsmethode zur Lösung des quantenmechanischen Mehrkörperprob-
   lems. Zeitschrift für Physik 61:126–148
4. Kohn W (1999) Nobel lecture: electronic structure of matter wave functions and density functionals. Rev Mod Phys 71:1253
5. Slater JC (1929) The theory of complex spectra. Phys Rev 34:1293
6. Löwdin P-O (1995) Quantum theory of many-particle systems. I. Physical interpretations by
   means of density matrices, natural spin-orbitals, and convergence problems in the method of
   configurational interaction. Phys Rev 97:1474
7. Hohenberg P, Kohn W (1964) Inhomogeneous electron gas. Phys Rev 136:B864
8. Bloch F (1929) Über die quantenmechanik der elektronen in kristallgittern. Zeitschrift für
   physik 52:555–600

9. Dirac PA(1930) Note on exchange phenomena in the Thomas atom. Mathematical proceedings of the Cambridge philosophical society, pp 376–385

10. Ceperley DM, Alder B (1980) Ground state of the electron gas by a stochastic method. Phys Rev Lett 45:566

11. Vosko S, Wilk L, Nusair M (1980) Accurate spin-dependent electron liquid correlation energies for local spin density calculations: a critical analysis. Can J Phys 58:1200–1211

12. Perdew JP, Yue W (1986) Accurate and simple density functional for the electronic exchange energy: generalized gradient approximation. Phys Rev B 33:8800

13. Perdew J, Burke K, Ernzerhof M (1996) Generalized gradient approximation made simple. Phys Rev Lett 77:3865–3868

14. Csonka GI, Perdew JP, Ruzsinszky A, Philipsen PH, Lebègue S, Paier J, Vydrov OA, Ángyán JG (2009) Assessing the performance of recent density functionals for bulk solids. Phys Rev B 79:155107

15. Perdew JP, Ruzsinszky A, Csonka GI, Vydrov OA, Scuseria GE, Constantin LA, Zhou X, Burke K (2008) Restoring the density-gradient expansion for exchange in solids and surfaces. Phys Rev Lett 100:136406

16. Allen JP, Scanlon DO, Watson GW (2010) Electronic structure of mixed-valence silver oxide AgO from hybrid density-functional theory. Phys Rev B 81:161103

17. Godby R, Schlüter M, Sham L (1986) Accurate exchange-correlation potential for silicon and its discontinuity on addition of an electron. Phys Rev Let 56:2415

18. Adamo C, Barone V (1999) Toward reliable density functional methods without adjustable parameters: the PBE0 model. J Chem Phys 110:6158

19. Perdew JP, Ernzerhof M, Burke K (1996) Rationale for mixing exact exchange with density functional approximations. J Chem Phys 105:9982–9985

20. Heyd J, Scuseria GE, Ernzerhof M (2003) Hybrid functionals based on a screened Coulomb potential. J Chem Phys 118:8207–8215

21. Heyd J, Scuseria GE, Ernzerhof M (2006) Erratum: "Hybrid functionals based on a screened Coulomb potential. J Chem Phys 24:219906 (J Chem Phys 118:8207, 2003)

22. Heyd J, Scuseria GE (2004) Efficient hybrid density functional calculations in solids: Assessment of the Heyd-Scuseria-Ernzerhof screened Coulomb hybrid functional. J Chem Phys 121:1187–1192

23. Grimme S (2011) Density functional theory with London dispersion corrections. WIREs Comput Mol Sci 1:211–228

24. Grimme S (2006) Semiempirical GGA-type density functional constructed with a long-range dispersion correction. J Comput Chem 27:1787–1799

25. Tkatchenko A, Scheffler M (2009) Accurate molecular van der Waals interactions from ground-state electron density and free-atom reference data. Phys Rev Lett 102:073005

26. Grimme S, Antony J, Ehrlich S, Krieg H (2010) A consistent and accurate ab initio parametrization of density functional dispersion correction (DFT-D) for the 94 elements H-Pu. J Chem Phys 132:154104

27. Tkatchenko A, DiStasio RA Jr, Car R, Scheffler M (2012) Accurate and efficient method for many-body van der Waals interactions. Phys Rev Lett 108:236402

28. Ambrosetti A, Reilly AM, DiStasio RA Jr, Tkatchenko A (2014) Long-range correlation energy calculated from coupled atomic response functions. J Chem Phys 140:18A508

29. Wimmer E, Freeman A, Weinert M, Krakauer H, Hiskes J, Karo A (1982) Cesiation of W (001): work function lowering by multiple dipole formation. Phys Rev Lett 48:1128

30. King-Smith R, Vanderbilt D (1993) Theory of polarization of crystalline solids. Phys Rev B 47:1651

31. Blöchl PE, Jepsen O, Andersen OK (1994) Improved tetrahedron method for Brillouin-zone integrations. Phys Rev B 49:16223

32. Lejaeghere K, Van Speybroeck V, Van Oost G, Cottenier S (2014) Error estimates for solid-state density-functional theory predictions: an overview by means of the ground-state elemental crystals. Crit Rev Solid State Mater Sci 39:1–24

Chapter 3
Computational Methodology

3.1 Structure Optimisation

The potential produced by any known exchange–correlation functional will always deviate from the true potential. As such, the ground state bond lengths and geometry of a material will not match those determined from experiment, even if measured at 0 K. It is therefore considered good practice to optimise a structure before further calculations are performed. For solid-state materials, both the positions of the nuclei and the lattice parameters must be converged. This is achieved by

1. calculating the forces acting on each atom,
2. perturbing the atoms to reduce the residual force, and
3. repeating the above until the forces fall within a certain tolerance.

The simplest technique for evaluating the atomic forces involves displacing each nuclei in all directions to find the force numerically. It is easy to see, however, that for large systems this approach will quickly become impractical. Instead, the Hellmann–Feynman theorem is employed [1], which states that if the distribution of the electrons is known, the atomic forces may be obtained from classical electrostatics. This enables the forces to be calculated from the wavefunction directly. The Hellmann–Feynman theorem can also be applied in calculating the stress tensor, thereby allowing for relaxation of the unit cell shape and volume. The quasi-Newton Raphson RMM-DIIS and conjugate gradient algorithms [2] were used to perturb the nuclei to their ground state positions once the atomic forces had been obtained.

This approach to complete structural optimisation leads to the unwanted introduction of *Pulay stress* [3]. When the volume of the lattice changes during relaxation, the plane wave basis set is no longer complete, resulting small errors in the diagonal components of the stress tensor. Pulay stress can be avoided by running a series of fixed-volume optimisations and fitting the energies to an equation of state to obtain the equilibrium volume. Alternatively, the plane wave energy cutoff can be raised significantly above that required to converge the total energy, at which point the Pulay

A. Ganose, *Atomic-Scale Insights into Emergent Photovoltaic Absorbers*, Springer Theses, https://doi.org/10.1007/978-3-030-55708-9_3

stress becomes negligible. In this thesis we have employed the latter approach, with the converged cutoff energy increased by ~30% for geometry optimisations.

3.2   Electronic Structure Methods

With the ground state geometry obtained, the electronic structure can then be characterised using a number of methods. Many of these are useful for assessing a compound's suitability as a photovoltaic absorber.

3.2.1   Density of States

The density of states (DOS) is a measure of the number of electronic states (per unit volume of a system) as a function of energy. The density of states can be compared with experimental spectra obtained from techniques such as X-ray photoelectron spectroscopy (XPS) and hard X-ray photoelectron spectroscopy (HAXPES). The total density of states can be decomposed into the contributions from individual atoms or orbitals, termed the *projected density of states*. Here, the wavefunction is projected onto the spherical harmonics for each atom within a certain radius—in our calculations defined by the PAW projection radii. In this thesis, all density of states spectra have been broadened to improve their legibility. This was achieved via convolution with a gaussian function with a width of 0.2 eV. In this thesis, density of states were plotted using the SUMO package developed by Ganose and Jackson [4].

3.2.2   Band Structure

Band structure diagrams are a plot of the electronic eigenvalues as a function of wave vector ($\mathbf{k}$). The band structure provides crucial information about the electronic properties of a system. For example, in semiconductors and insulators, analysis of the frontier valence and conduction bands indicates the magnitude and nature (direct vs. indirect) of the fundamental band gap. In order to reduce the computational cost, it is essential to choose a path through $\mathbf{k}$-space that will be representative of the reciprocal lattice and capture as many interesting phenomena as possible. A common technique is to plot a path along the high-symmetry lines of the first Brillouin zone. These paths are unique for each crystal class. We have used the coordinates of the high-symmetry points defined in Bradley and Cracknell [5]. Band structures were constructed and plotted using the SUMO package [4].

3.2.3 Carrier Effective Mass

The band structure of a material also provides information about the effective mass of charge carriers. In particular, bands displaying a greater degree of dispersion at the band edges give rise to charge carriers with smaller effective mass, in comparison to flat bands which indicate localised states. The effective mass of a charge carrier, m^*, at the band edges is calculated as

$$\frac{\hbar}{m^*} = \frac{d^2 E(\mathbf{k})}{d\mathbf{k}^2},$$

(3.1)

where $d^2 E(\mathbf{k})/d\mathbf{k}^2$ is the curvature of the band at the band edge. Anisotropy in the band curvature was taken into account through parabolic fitting of the band in all high symmetry directions. The relationship between effective mass and carrier mobility, μ, is given by

$$\mu = \frac{e\tau}{m^*},$$

(3.2)

where e is the elementary charge and τ is the average scattering time. The calculation of τ is computationally demanding since it depends on the doping and defect properties of a system. Accordingly, in this thesis we provide the effective mass as an indicator of *potential* mobility. Effective masses were calculated using the SUMO package [4].

3.2.4 Band Alignment

A limitation of periodic electronic structure calculations is the absence of any external reference energy. In order to align the electronic eigenvalues to the vacuum level—and thus calculate the ionisation potential and electron affinity—further calculations are required. In this thesis, we have adopted the core-level alignment approach of Wei and Zunger [6], whereby:

1. A slab model is constructed containing a significantly large vacuum region such that periodic images do not interact.
2. The electrostatic potential is averaged in the direction perpendicular to the surface of the slab. The potential in the middle of the vacuum region is taken as the vacuum level (E_{vac}).
3. The energy of a core state deep in the slab is taken as a reference (E_{core}^{slab}). The slab should be thick enough that this reference state is a good approximation for "bulk-like" behaviour.
4. The reference energies are compared against the valence band maximum (E_{vbm}) and an equivalent core energy (E_{core}^{bulk}) taken from a calculation on the bulk material.

The ionisation potential (IP) can then be calculated according to

$$IP = \left(E_{\text{vac}} - E_{\text{core}}^{\text{slab}}\right) - \left(E_{\text{vbm}} - E_{\text{core}}^{\text{bulk}}\right). \tag{3.3}$$

The planar average of the calculated electrostatic potential was extracted and plotted using the MACRODENSITY package [7–10].

3.2.5 Crystal Orbital Hamilton Population

Crystal orbital Hamilton population (COHP) analysis provides information on the strength and nature of bonding in a system. By rewriting the total energy as a sum of orbital pair contributions, an energy resolved COHP diagram can be produced—effectively a partitioned density of states. The sign of the COHP indicates the bonding type—with positive and negative values indicating bonding and anti-bonding, respectively—whereas the magnitude indicates the bond strength, with larger values indicating stronger interactions.

The COHP method was originally proposed by Blöchl to aid the analysis of bonding in periodic calculations employing a localised basis set [11]. Recently [12], showed that by projecting the wavefunction onto a set of localised orbitals, $\{\phi_\mu\}$, the COHP method could be extended for use in plane wave calculations. Here, the transfer matrix, $\mathbf{T}$, quantifies the overlap between the band wavefunctions, ψ_j, and the local orbitals, as

$$T_{j\mu}(\mathbf{k}) = \langle \psi_j(\mathbf{k})|\phi_\mu\rangle. \tag{3.4}$$

As the COHP is given between pairs of orbitals, the aim is to calculate the bonding strength between an orbital, μ, located at the first atom involved in the bond in question, and another orbital, ν, on the second atom. To achieve this, the projected density matrix, $\mathbf{P}^{\text{proj}}$, is defined as

$$P_{j\mu\nu}^{\text{proj}}(\mathbf{k}) = T_{j\mu}^\dagger(\mathbf{k})T_{j\nu}(\mathbf{k}). \tag{3.5}$$

The Hamiltonian must be expressed in the basis of the local functions [13]. This is achieved by expanding the plane wave Hamiltonian, $\hat{H}^{\text{pw}}$, according to

$$\begin{aligned}\hat{H}_{\mu\nu}^{\text{proj}}(\mathbf{k}) &= \langle \phi_\mu|\hat{H}^{\text{pw}}|\phi_\nu|\rangle \\ &= \sum_j \langle \phi_\mu|\psi_j(\mathbf{k})\rangle \varepsilon_j(\mathbf{k})\langle \psi_j(\mathbf{k})|\phi_\nu\rangle.\end{aligned} \tag{3.6}$$

Using Eq. 3.4 we can write

$$\hat{H}_{\mu\nu}^{\text{proj}}(\mathbf{k}) = \sum_j \varepsilon_j(\mathbf{k})T_{j\mu}^\dagger(\mathbf{k})T_{j\nu}(\mathbf{k}). \tag{3.7}$$

Finally, the energy-dependent projected COHP (pCOHP) can be calculated as

$$\text{pCOHP}_{\mu\nu}(E, \mathbf{k}) = \sum_j \text{Re} \left[P_{j\mu\nu}^{\text{proj}}(\mathbf{k}) \hat{H}_{\mu\nu}^{\text{proj}}(\mathbf{k}) \right] \times \delta(\varepsilon_j(\mathbf{k}) - E), \qquad (3.8)$$

where δ is the Dirac delta function. Accordingly, the projected density matrix has been transformed into a density-of-states matrix. The real-space pCOHP(E) is calculated by summing over all orbitals (μ and ν) and integrating over $\mathbf{k}$-space.

In this report, pCOHP analysis was performed using the LOBSTER program, in which the methodology above has been adapted for use in the PAW formalism [14, 15]. The tetrahedron method was employed when integrating over $\mathbf{k}$-space [16, 17]. The remaining challenge is in choosing a suitable basis set of localised functions. LOBSTER is provided with a minimal basis set of Slater-type orbitals, fitted to atomic wavefunctions, that possess the correct nodal behaviour in the core region [18, 19]. The basis set further provides additional basis functions for unoccupied atomic orbitals based on fitting to plane wave DFT free-atom calculations [15].

3.3 Dielectric Response

The dielectric response quantifies a material's ability to screen charge. The dielectric response at low-frequencies is termed the static dielectric constant. A large static dielectric constant is thought to improve defect tolerance by screening the effects of charged defects on carrier transport and non-radiative recombination. The static dielectric constant is comprised of contributions from both the electron density and the lattice.

3.3.1 High-Frequency Dielectric Spectra

The electronic component is obtained in the calculation of the high-frequency dielectric spectra. Obtaining the high-frequency dielectric response of a material is also crucial for understanding its optical properties. The process by which this is achieved within the PAW formalism is not trivial and has been addressed in detail in [20] Here, the imaginary part of the dielectric constant (ε_r'') is obtained by a summation over all direct valence band to conduction band transitions. Accordingly, this method fails to account for indirect and intraband transitions. The real part of the dielectric response (ε_r') is then obtained from the imaginary part by the Kramers-Kronig transformation. This single-particle approach does not include electron–hole correlations and other many-body effects. Accordingly, for very accurate descriptions of the dielectric response, higher order methods such as the Bethe-Salpeter equation should be employed [21]. Regardless, this approach has previously been shown to provide reasonable agreement with experiment [22].

3.3.2   Ionic Contribution to the Dielectric Constant

The lattice contribution depends on the Born effective charges—effectively a measure
of how much charge follows an atom when it is displaced—and the phonon modes at
the Γ point. These properties can be obtained either through finite displacements or
density functional perturbation theory (DFPT), with the methodology behind both
approaches covered in detail in [23–25]. In thesis we have employed the DFPT
approach due to the reduced computational cost.

3.4   Point Defects

The presence of defects is the primary factor controlling conductivity and recombination in solar absorbers. Defects are usually distributed randomly throughout a
material and are often present at low concentrations. To model the effects of point
defects in solid-state calculations, we have employed the *supercell approximation*.
Here, a defect is introduced into an expansion of the unit cell containing multiple
primitive cells (termed a supercell), with the aim of simulating a defect *in solution*.
Crucially, it is essential to minimise the interaction between a defect and its images in
neighbouring cells. In practice, the supercell size is limited based on the availability
of computational resources.

3.4.1   Defect Formation Energy

The formation energy, $\Delta_f H$, of a defect, X, with charge state q, is calculated as

$$\Delta_f H^{X,q} = \left(E^{X,q} - E^H\right) + \sum_i [n_i (E_i + \mu_i)] + q \left(E_F + \varepsilon_{vbm}^H\right) + E_{corr}, \quad (3.9)$$

where $E^{X,q}$ is the energy of the defected supercell and E^H is the energy of the
unperturbed host supercell. The second term represents the energy change due to
losing an atom, i, to a chemical reservoir: n_i is the number of atoms of each type
lost, E_i is the element reference energy calculated from the element in its standard
state, and μ_i is the chemical potential of the atom, which can be used to explore typical
growth conditions. The first two terms can be understood simply as an expression of
Hess' Law—i.e. the formation energy is effectively the energy of the products minus
the reactants. The third term is only required when calculating the formation energy
of a charged defect and accounts for the exchange of charge carriers with a carrier
reservoir: ε_{vbm}^H represents the energy needed to add or remove an electron from the
valence band maximum to a Fermi reservoir—i.e. the eigenvalue of the valence band
maximum in the host—and E_F is the Fermi level relative to ε_{vbm}^H. E_{corr} is a correction

applied to account for various limitations of the defect scheme used and is comprised
of three terms,

$$E_{\text{corr}} = E_{\text{pot}} + E_{\text{bf}} + E_{\text{icc}}, \tag{3.10}$$

where E_{pot} is a correction to account for potential alignment mismatch, and E_{bf} and
E_{icc} are corrections to account for finite supercell effects.

3.4.1.1 Potential Alignment Correction

As previously mentioned, in periodic density functional theory, the total energy of a
system is calculated relative to the background electrostatic potential. When calculating charged defects, the removal or addition of an electron requires the introduction
of a jellium background that neutralises the overall charge of the cell. Accordingly,
there is a mismatch between the reference electrostatic potential of the charged defect
compared to that of the neutral host supercell. A correction must therefore be applied
to ensure the total energies of both systems are comparable. This is defined as

$$E_{\text{pot}}^{X,q} = q \left[V_{\text{r}}^{X,q} - V_{\text{r}}^{\text{H}} \right], \tag{3.11}$$

where V_{r}^{H} is the potential at a reference point in the host and $V_{\text{r}}^{X,q}$ is the potential at
the same reference point in the defected supercell. In practise, a core level far from
the defect site is chosen as the reference point.

3.4.1.2 Band Filling Correction

Due to finite supercell sizes, the introduction of a defect causes an impurity–impurity
interaction resulting in the formation of a defect band. In an infinitely large supercell,
the defect would instead result in a single defect eigenstate. If the defect band lies
deep within the band gap it will likely remain localised and show minimal effect on
the surrounding electronic structure. In contrast, if the band lies within or close to
the band edges, any electrons or holes that are introduced will artificially occupy the
valence or conduction bands according to a Fermi–Dirac distribution. A correction
is therefore applied to account for the effects on the total energy of inadvertent *band
filling* [26]. For shallow donors the correction is calculated as

$$E_{\text{bf}} = - \sum_{j,\mathbf{k}} \Theta \left(\varepsilon_{j,\mathbf{k}} - \tilde{\varepsilon}_{\text{cbm}} \right) w_{\mathbf{k}} \eta_{j,\mathbf{k}} \left(\varepsilon_{j,\mathbf{k}} - \tilde{\varepsilon}_{\text{cbm}} \right), \tag{3.12}$$

with the correction for shallow acceptors given as

$$E_{\text{bf}} = \sum_{j,\mathbf{k}} \Theta \left(\tilde{\varepsilon}_{\text{vbm}} - \varepsilon_{j,\mathbf{k}} \right) w_{\mathbf{k}} \left(1 - \eta_{j,\mathbf{k}} \right) \left(\varepsilon_{j,\mathbf{k}} - \tilde{\varepsilon}_{\text{vbm}} \right), \tag{3.13}$$

where $\varepsilon_{j,\mathbf{k}}$ are electronic eigenvalues from the defect calculation, $\tilde{\varepsilon}_{\mathrm{cbm}}$ and $\tilde{\varepsilon}_{\mathrm{vbm}}$ are the eigenvalues of the conduction band minimum and valence band maximum of the host supercell after the potential alignment is applied, Θ is the Heaviside step function, $w_{\mathbf{k}}$ is the $\mathbf{k}$-point weight, and $\eta_{j,\mathbf{k}}$ is the band occupation.

3.4.1.3 Image-Charge Correction

Charged defects require a further correction due to the slow decay of the Coulomb interaction with distance. Generally, the size of the supercells used do not allow for this interaction to decay completely, resulting in the spurious electrostatic interaction between defect charge sites in neighbouring cells [27]. An *image-charge* correction is introduced to restore the energy of the system to that of a charged defect in the dilute limit. While several forms of this correction exist, most are based on evaluating the potential between two point charges in a neutralising jellium—the Madelung energy—calculated as

$$E_{\mathrm{Madelung}} = -\frac{q^2\alpha}{2L\varepsilon}, \tag{3.14}$$

where q is the charge, α is the structure-dependent Madelung constant, L is the distance between the charges, and ε is the static dielectric constant [28].

In practice, the electron density of a charged defect rarely acts as a point charge due to some degree of delocalisation. To account for this, [28] introduced a third-order expression taking into account the quadrupole moment of the charge density, Q, as

$$E_{\mathrm{icc}}^{\mathrm{MP}} = -\frac{q^2\alpha}{2L\varepsilon} - \frac{2\pi qQ}{3\varepsilon L^3}. \tag{3.15}$$

Subsequently, Lany and Zunger demonstrated that the second term of this expression scaled proportional to q^2/L, instead of the expected q/L^3, due to the effect of the dielectric screening on Q [29]. Accordingly, they derived an approximate expression for the correction based on the shape factor of the cell, c_{sh}, as

$$E_{\mathrm{icc}}^{\mathrm{LZ}} = \left[1 + c_{\mathrm{sh}}\left(1 - \varepsilon^{-1}\right)\right]\frac{q^2\alpha}{2L\varepsilon}. \tag{3.16}$$

For cases where ε is sufficiently large and the supercell is isotropic—as is the case for all materials studied in this thesis—the pre-factor can be further approximated as $2/3$. We have additionally used the formalism developed by Murphy et al., which takes into account any anisotropy in the dielectric screening [30].

3.4.2 Thermodynamic Transition Levels

The presence of a defect often introduces defect states into the band gap of a material. These can be measured experimentally through techniques such as deep-level transient spectroscopy and are used to assess the nature of the defect. These states do not correspond to the eigenvalues of the Kohn–Sham wavefunctions obtained directly from the calculation on the defected system. Instead, the states of interest generally involve the transition from one defect charge state to another—resulting in the release or capture of an electron or hole.

The energy at which the charge state of a defect spontaneously transforms from $q \leftrightarrow q'$ is termed the *thermodynamic transition level*, and is calculated as

$$\varepsilon(X, q/q') = \frac{\Delta_f H^{X,q} - \Delta_f H^{X,q'}}{q' - q}. \tag{3.17}$$

The position of the thermodynamic transition level relative to the valence and conduction band edges indicates whether a defect will contribute to conductivity or act as a charge trap. A defect state is termed shallow if it is within $k_B T$ of the band edges. In particular, shallow donor defects near the conduction band minimum facilitate *n*-type conductivity, whereas shallow acceptor defects close to the valence band maximum promote *p*-type conductivity.

3.4.3 Chemical Potential Limits

The chemical potentials required to calculate the defect formation energy reflect the chemical reservoirs for the species involved in the defect. In practice, these reservoirs are controlled by the experimental growth conditions. Accordingly, by varying the chemical potentials one can explore a range of different chemical environments. The range of accessible chemical potentials is limited by the formation of secondary phases. As such, one must also calculate the total energy of all competing phases to ensure the set of chemical potentials remains experimentally achievable.

The chemical potentials are calculated by imposing a series of bounds. To illustrate this process we take GaAs as an example. Initially, the chemical potentials of gallium and arsenic are related by the stability of the GaAs phase, namely

$$\mu_{Ga} + \mu_{As} = \mu_{GaAs} = \Delta_f H^{GaAs}. \tag{3.18}$$

The chemical potentials are limited by the formation of elemental gallium and arsenic as

$$\mu_{Ga} \leq \mu_{Ga}^{elemental}, \tag{3.19}$$

$$\mu_{As} \leq \mu_{As}^{elemental}, \tag{3.20}$$

where

$$\mu_{Ga}^{elemental} = \Delta_f H^{Ga} = 0, \tag{3.21}$$

$$\mu_{As}^{elemental} = \Delta_f H^{As} = 0. \tag{3.22}$$

In addition, the lower limit of μ_{Ga}, indicating a gallium poor environment, is dictated by the formation of elemental arsenic—e.g. by setting $\mu_{As} = 0$ in Eq. 3.18. Analogously, μ_{As} is limited by the formation of elemental gallium. Together this gives

$$\mu_{Ga} \geq \Delta_f H^{GaAs}, \tag{3.23}$$

$$\mu_{As} \geq \Delta_f H^{GaAs}. \tag{3.24}$$

Solving this set of simultaneous equations gives the range of thermodynamically accessible chemical potential environments. The CPLAP code was employed to calculate the chemical potential limits of the ternary systems studied in this thesis [31].

3.4.4 Transition Level Diagrams

The defect properties of a system are commonly presented in a *transition level diagram* in which the defect formation energies are plotted as a function of Fermi level, at a fixed set of chemical potentials. Accordingly, information regarding both the thermodynamic cost and electronic behaviour of a defect is conveyed simultaneously. The transition level diagram for multiple chemical potential environments are often displayed together to indicate how the defect behaviour varies with experimental growth conditions. While a defect may possess many charge states, only the most stable charge state at each Fermi level is indicated. For example, in the case of the anion vacancy shown in Fig. 3.1a, the +1 charge state is higher in energy than the neutral and +2 charge states at all Fermi levels and will therefore not be seen experimentally.

In these diagrams, the slope of the line indicates the charge state of the defect, with donor defects—i.e. those donating one or more electrons to the system—sloping upwards from left to right and acceptor defects—those donating one or more holes—sloping downwards. Neutral defects are indicated by horizontal lines, as their formation energies do not depend on the position of the Fermi level. The transition levels—also termed ionisation levels—are indicated by filled circles. The Fermi level is pinned *roughly* at the position where the lowest energy donor and acceptor defects cross, as indicated in Fig. 3.1b. Accordingly, as the chemical potentials are adjusted the position of the Fermi level can be tuned.

Fig. 3.1 Example transition level diagrams for a hypothetical material. **a** Only the lowest energy charge state at each Fermi level of the anion vacancy is indicated by the solid red line. **b** The Fermi level is pinned near where the lowest energy donor and acceptor defects cross

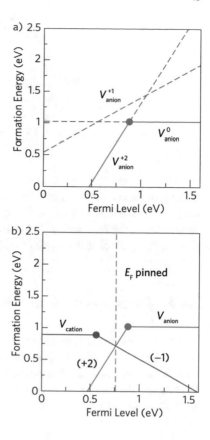

3.4.5 Self-consistent Fermi Level

While the position of the Fermi level can be estimated based on visual inspection of the transition level diagram as described above, in practice E_F will depend on multiple factors. These include the density of states and temperature of the sample. The Fermi level can be obtained in a self-consistent manner based on a set of thermodynamic transition levels, given the condition of charge neutrality [32]

$$n_0 + \sum_i |q_i| N_{q_i}^A = p_0 + \sum_j |q_j| N_{q_j}^D, \qquad (3.25)$$

where n_0 and p_0 are the concentration of electrons and holes respectively, $N_{q_i}^A$ is the concentration of acceptor defects in charge state q_i, and $N_{q_j}^D$ is the concentration of donor defects in charge state q_j. These concentrations depend on E_F as

$$N^{X,q} = N_0 \exp \left[\frac{\Delta_f H^{X,q}(E_F)}{k_B T} \right],$$ (3.26)

$$n_0 = \int_0^\infty \frac{1}{e^{(E - E_F)/k_B T} + 1} g(E) \, dE,$$ (3.27)

$$p_0 = \int_0^\infty \left[1 - \frac{1}{e^{(E - E_F)/k_B T} + 1} \right] g(E) \, dE,$$ (3.28)

where N_0 is the density of sites where defect X can form and $g(E)$ is the bulk density of states. Accordingly, the occupation of the valence and conduction bands (e.g. concentrations of holes and electrons) is determined by Fermi–Dirac statistics. In this thesis, the self-consistent Fermi level was calculated using the SC- FERMI code developed by Buckeridge et al. [33–35].

3.4.6 Shockley–Read–Hall Recombination Rate

The defect transition levels also play a role in determining the rate of Shockley–Read–Hall (trap-assisted) recombination [36]. Here, the rate of recombination is determined by quasi-equilibrium semiconductor statistics. Crucially, in steady state the recombination rate of electrons and holes must be equal. Within this model, the recombination rate, R^{SRH}, is calculated as

$$R^{SRH} = \frac{(n + \Delta n)(p + \Delta p) - n_i^2}{\tau_n(n + \Delta n + n_1) + \tau_p(p + \Delta p + p_1)},$$ (3.29)

where p and n are the self-consistent hole and electron concentrations, Δp and Δn are the change in hole and electron concentrations resulting from steady-state illumination and were set as $1 \times 10^{14} \text{cm}^{-3}$, and n_i is the concentration of electrons in an intrinsic sample. τ_p (τ_n) is the lifetime of holes (electrons) in the limit that all recombination centres are occupied with electrons (holes), also termed the Shockley–Read–Hall lifetime. The lifetimes are calculated as

$$\tau = 1/N_X \nu \sigma,$$ (3.30)

where N_X is the concentration of the defect, ν is the group velocity, and σ is the defect capture cross-section. Due to the immense computational difficulty in calculating capture cross-sections we have assumed a constant value for all defects.

Returning to Eq. 3.29, p_1 and n_1 indicate the concentration of holes and electrons if the Fermi level is pinned at the same energy as the defect transition level, ε_X. These are calculated according to Fermi–Dirac statistics as

$$p_1 = N_v \exp\left[\frac{-\varepsilon_X - \varepsilon_{vbm}}{k_B T}\right], \tag{3.31}$$

$$n_1 = N_c \exp\left[\frac{\varepsilon_X - \varepsilon_{cbm}}{k_B T}\right], \tag{3.32}$$

$$n_i^2 = n_1 \times p_1 \tag{3.33}$$

where N_v and N_c are the effective density of states of the valence and conduction bands, respectively, and are defined as

$$N_v = 2\left(\frac{m_h^* k_B T}{2\pi \hbar^2}\right)^{3/2}, \tag{3.34}$$

$$N_c = 2\left(\frac{m_e^* k_B T}{2\pi \hbar^2}\right)^{3/2}. \tag{3.35}$$

3.5 Vienna *Ab Initio* Simulation Package

All calculations in this thesis were carried out in the framework of density functional theory, using the Vienna *Ab initio* Simulation package (VASP) [37–40]. VASP is a periodic plane wave code, employing the projector augmented-wave method to describe the interactions between core and valence electrons [41]. Due the complete set of PAW pseudopotentials included in the distribution and the high level of parallelisability, VASP is employed widely among the computational chemistry and physics communities. The VESTA package was used to plot crystal structures and charge density isosurfaces directly from the output of VASP calculations [42].

References

1. Feynman RP (1939) Forces in molecules. Phys Rev 56:340
2. Pulay P (1980) Convergence acceleration of iterative sequences. The case of SCF iteration. Chem Phys Lett 73:393–398
3. Pulay P (1969) Ab initio calculation of force constants and equilibrium geometries in polyatomic molecules: I. Theory Mol Phys 17:197–204
4. https://github.com/SMTG-UCL/sumo. Accessed 14 Mar 2018
5. Bradley C, Cracknell AP (1972) The mathematical theory of symmetry in solids: representation theory for points groups and space groups. Clarendon Press
6. Wei S-H, Zunger A (1998) Calculated natural band offsets of all II-VI and III-V semiconductors: chemical trends and the role of cation d orbitals. Appl Phys Lett 72:2011–2013
7. Yang Z, Chueh C-C, Zuo F, Kim JH, Liang P-W, Jen AK-Y (2015) High-performance fully printable perovskite solar cells via blade-coating technique under the ambient condition. Adv Energy Mater 5:1500328
8. Walsh A, Butler KT (2013) Prediction of electron energies in metal oxides. Acc Chem Res 47:364–372

9. Burton LA, Walsh A (2013) Band alignment in SnS thin-film solar cells: possible origin of the low conversion efficiency. Appl Phys Lett 102:132111

10. https://github.com/WMD-group/MacroDensity. Accessed 14 Mar 2018

11. Dronskowski R, Blöchl PE (1993) Crystal orbital hamilton populations (COHP): energy-resolved visualization of chemical bonding in solids Based on Density-Functional Calculations. J Phys Chem 97:8617–8624

12. Deringer VL, Tchougréeff AL, Dronskowski R (2011) Crystal orbital Hamilton population (COHP) analysis as projected from plane-wave basis sets. J Phys Chem A 115:5461–5466

13. Sanchez-Portal D, Artacho E, Soler JM (1995) Projection of plane-wave calculations into atomic orbitals. Solid State Commun 95:685–690

14. Maintz S, Deringer VL, Tchougréeff AL, Dronskowski R (2013) Analytic projection from plane-wave and paw wavefunctions and application to chemical-bonding analysis in solids. J Comput Chem 34:2557–2567

15. Maintz S, Deringer VL, Tchougréeff AL, Dronskowski R (2016) LOBSTER: a tool to extract chemical bonding from plane-wave based DFT. J Comput Chem 37:1030–1035

16. Jepsen O, Andersen O (1984) No error in the tetrahedron integration scheme. Phys Rev B 29:5965

17. Blöchl PE, Jepsen O, Andersen OK (1994) Improved tetrahedron method for Brillouin-zone integrations. Phys Rev B 49:16223

18. Bunge CF, Barrientos JA, Bunge AV (1993) Roothaan-Hartree-Fock ground-state atomic wave functions: slater-type orbital expansions and expectation values for Z= 2–54. At Data Nucl Data Tables 53:113–162

19. Koga T, Kanayama K, Watanabe S, Thakkar AJ (1999) Analytical Hartree-Fock wave functions subject to cusp and asymptotic constraints: He to Xe, Li^+ to Cs^+, H^- to I^-. Int J Quantum Chem 71:491–497

20. Gajdoš M, Hummer K, Kresse G, Furthmüller J, Bechstedt F (2006) Linear optical properties in the projector-augmented wave methodology. Phys Rev B 73:045112

21. Sham L, Rice T (1966) Many-particle derivation of the effective-mass equation for the Wannier exciton. Phys Rev 144:708

22. Birkett M, Savory CN, Fioretti AN, Thompson P, Muryn CA, Weerakkody A, Mitrovic I, Hall S, Treharne R, Dhanak VR, Scanlon DO, Zakutayev A, Veal TD (2017) Atypically small temperature-dependence of the direct band gap in the metastable semiconductor copper nitride Cu_3N. Phys Rev B 95:115201

23. Baroni S, Resta R (1986) Ab initio calculation of the macroscopic dielectric constant in silicon. Phys Rev B 33:7017

24. Gonze X, Lee C (1997) Dynamical matrices, Born effective charges, dielectric permittivity tensors, and interatomic force constants from density-functional perturbation theory. Phys Rev B 55:10355

25. Nunes R, Gonze X (2001) Berry-phase treatment of the homogeneous electric field perturbation in insulators. Phys Rev B 63:155107

26. Lany S, Zunger A (2008) Assessment of correction methods for the band-gap problem and for finite-size effects in supercell defect calculations: case studies for ZnO and GaAs. Phys Rev B 78:235104

27. Freysoldt C, Neugebauer J, Van de Walle CG (2009) Fully ab initio finite-size corrections for charged-defect supercell calculations. Phys Rev Lett 102:016402

28. Makov G, Payne M (1995) Periodic boundary conditions in ab initio calculations. Phys Rev B 51:4014

29. Lany S, Zunger A (2009) Accurate prediction of defect properties in density functional supercell calculations. Modell Simul Mater Sci Eng 17:084002

30. Murphy ST, Hine ND (2013) Anisotropic charge screening and supercell size convergence of defect formation energies. Phys Rev B 87:094111

31. Buckeridge J, Scanlon D, Walsh A, Catlow CRA (2014) Automated procedure to determine the thermodynamic stability of a material and the range of chemical potentials necessary for its formation relative to competing phases and compounds. Comput Phys Commun 185:330–338

32. Kittel C, Kroemer H, Scott H (1980) Thermal physics, 2nd edn. Freeman and Co., New York (Chap. 13)
33. Taylor FH, Buckeridge J, Catlow CRA (2016) Defects and oxide ion migration in the solid oxide fuel cell cathode material $LaFeO_3$. Chem Mater 28:8210–8220
34. Buckeridge J, Jevdokimovs D, Catlow C, Sokol A (2016) Nonstoichiometry and Weyl fermionic behavior in TaAs. Phys Rev.B 94:180101
35. https://github.com/jbuckeridge/sc-fermi. Accessed 14 Mar 2018
36. Shockley W, Read W Jr (1952) Statistics of the recombinations of holes and electrons. Phys Rev 87:835
37. Kresse G, Hafner J (1993) Ab initio molecular dynamics for liquid metals. Phys Rev B 47:558–561
38. Kresse G, Hafner J (1994) Ab initio molecular-dynamics simulation of the liquid-metal amorphous-semiconductor transition in germanium. Phys Rev B 49:14251–14269
39. Kresse G, Furthmüller J (1996) Efficient iterative schemes for ab initio total-energy calculations using a plane-wave basis set. Phys Rev B 54:11169–11186
40. Kresse G, Furthmüller J (1996) Efficiency of ab initio total energy calculations for metals and semiconductors using a plane wave basis set. Comput Mater Sci 6:15
41. Kresse G, Joubert D (1999) From ultrasoft pseudopotentials to the projector augmented-wave method. Phys Rev B 59:1758
42. Momma K, Izumi F (2008) VESTA: a three-dimensional visualization system for electronic and structural analysis. J Appl Crystallogr 41:653–658

Part II
Perovskite-Inspired Absorbers

Chapter 4
Review: Perovskite Photovoltaics

As previously discussed, $CH_3NH_3PbI_3$ (MAPI) and the hybrid perovskites have recently emerged as a remarkably efficient class of solar absorbers. Unfortunately, MAPI suffers from several issues that limit its commercial viability. A primary issue is device hysteresis, whereby significant differences are observed for $J-V$ curves measured under forward and reverse bias. Planar MAPI devices containing thin mesoporous TiO_2 and small grain sizes are particularly affected and show losses in fill factors and efficiencies of up to 33% [1]. The presence of hysteresis has been ascribed to high levels of ionic conduction [2, 3]. In particular, mobile methylammonium [4] and iodine vacancies [5] that can migrate throughout the absorber layer—in response to the photogenerated voltage—are thought to generate an electric field that opposes that of the $p-n$ junction.

While recent device improvements have minimised the effect of hysteresis [6, 7], the stability of the hybrid perovskites is a critical issue which presents a major roadblock in the move toward commercialisation. Stability is particularly crucial as device longevity is essential to reach energy payback times [8]. The facile decomposition of MAPI in moisture has been known since the earliest hybrid perovskite cells were produced [9, 10], with further studies detailing the hydrolysation of MAPI upon contact with air [11–13]. The thermal stability of MAPI is also poor, with decomposition occurring at temperatures above 85 °C [14, 15]. Recently, several studies, both theoretical and experimental, have indicated that MAPI is intrinsically unstable with respect to phase separation into PbI_2 and $CH_3NH_3I_3$ due to its small formation energy [16–19]. As such, modifications to the MAPI structure that confer greater stability are highly desirable [20].

4.1 Perovskite Structure

MAPI adopts the perovskite ABX_3 structure (where A = CH_3NH_3, B = Pb, and X = I), where corner sharing PbI_6 octahedra form a cage enclosing the organic cation (Fig. 4.1). One route to tuning the properties of MAPI is through changing the non-

© The Editor(s) (if applicable) and The Author(s), under exclusive license
to Springer Nature Switzerland AG 2020
A. Ganose, *Atomic-Scale Insights into Emergent Photovoltaic Absorbers*,
Springer Theses, https://doi.org/10.1007/978-3-030-55708-9_4

Fig. 4.1 Schematic of
perovskite structure
indicating the A, B, and X
lattice sites

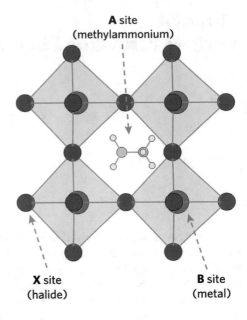

A site
(methylammonium)

X site
(halide)

B site
(metal)

bonding organic component. To date, after over nine years of development, the only
organic cations to be successfully incorporated into the perovskite motif are methylammonium (MA) and formamidinium (FA). Incorporation of larger cations results
in reduced dimensionality structures due to disruption of the three-dimensional (3D)
lead iodide cage [21–23]. Introducing FA in the place of MA to form $CH(NH_2)_2PbI_3$
(FAPI), produces films with a slightly reduced band gap of 1.48 eV, improved PCEs,
extended photoluminescence (PL) lifetimes, and greater thermal stability [24, 25].
Unfortunately, FAPI can also form in a thermally accessible (less than 360 K) hexagonal δ-phase possessing a large band gap, which cannibalises device performance
[26, 27]. Including 20% MA in the synthesis of FAPI stabilises the black α-phase
whilst preserving long exciton lifetimes and high efficiencies [28, 29], however, the
long-term stability of these devices is yet to be addressed.

An alternative method for improving long-term stabilities is to completely replace
the organic component, forming an all-inorganic perovskite [30, 31]. Eperon et al.
recently produced a functioning $CsPbI_3$ device through suppression of the formation of a competing non-perovskite yellow phase during synthesis [32, 33]. Despite
remarkable thermal stability up to 300 °C, their films were highly sensitive to ambient
conditions and performed poorly, reaching a PCE of only 2.9%.

MAPI can also be tuned on the X site by replacing iodine with other halides
[34]. As the position of the valence band maximum is largely controlled by the
binding energy of the halide valence p orbitals, swapping I for Br and Cl results in
materials with band gaps too large for photovoltaic applications. However, mixed-
halide perovskites are commonly used to fine-tune the absorber properties. In the
first perovskite cells, $PbCl_2$ used as a precursor resulted in a small proportion of

the chlorine being incorporated into the perovskite layer, in turn leading to more even distribution of nucleation sites and smoother films [6, 35, 36]. Furthermore, use of bromide in a solid solution allows for an adjustable band gap [37] and reduced levels of hysteresis [7]. Rehman et al. trialled a mixed bromide/iodide absorber layer, $FAPb(Br_xI_{1-x})_3$, as a potential top cell in tandem devices [38]. While the composition where $x = 0.3$–0.5 allowed for an ideal top cell band gap of $\sim$1.7–1.8 eV, the resulting material appeared amorphous with weak optical absorption and dramatically reduced charge-carrier diffusion lengths. This observation has been confirmed by theoretical calculations on the related $MAPb(Br_xI_{1-x})_3$ system, with the region between $0.3 < x < 0.6$ showing intrinsic instability with respect to spinodal decomposition at 300 K [39].

McMeekin et al. have recently demonstrated a mixed-cation mixed-halide perovskite system which subjugates the aforementioned phase instability region through partial substitution of formamidinium with caesium [40]. With a composition of $FA_{0.83}Cs_{0.17}Pb(I_{0.6}Br_{0.4})_3$, their highly crystalline films possessed a large V_{oc} of 1.2 eV and efficiencies up to 17.9%. When employed as a top cell coupled with a crystalline silicon module in a tandem device, the same composition achieved an efficiency of 19.8%. In 2017, inclusion of rubidium cations in a mixed-cation mixed-halide perovskite produced thin films with a band gap of 1.73 eV [41]. When combined with an interdigitated back-contact silicon cell, devices reached an impressive efficiency of 26.4%.

4.2 Reduced Dimensionality Perovskites

Over the last four years, layered perovskites have emerged as a promising route to increased stabilities [42, 43]. When large organic cations are incorporated in the synthesis of the hybrid perovskites, lower dimensionality structures are formed as the cations cannot fit within the perovskite cage [44, 45]. These structures, composed of layers of PbI_6 octahedra "capped" by organic molecules, form two-dimensional (2D) or Ruddlesden–Popper like phases [46]. The primary mode of stabilisation is through van der Waals interactions that occur between the capping molecules and the surface of the inorganic layer [47–50]. Analogous to MAPI, the physical and electronic properties of the layered perovskites can be tuned by modifying the organic, metal and halide components [21, 22]. The success of the layered perovskites is partially due to the ability to control the thickness of the inorganic layer. By tailoring the stoichiometric quantities of the organic versus inorganic components, the number of layers of perovskite octahedra per slab (n), for the series $(PEA)_2(CH_3NH_3)_{n-1}Pb_nI_{3n+1}$ (where $PEA = C_6H_5(CH_2)_2NH_3^+$), can be regulated [51]. In this scheme, the limit $n = \infty$ corresponds to the 3D perovskite structure, $n = 1$ denotes the 2D layered structure, and $n > 1$ describes "quasi-2D" systems (Fig. 4.2).

In the 1990s, Mitzi et al. characterised a series of intercalated organic–inorganic germanium, tin, and lead perovskites, noting the role of large organic cations in promoting enhanced thermal and chemical stability. properties [46–48, 51–57]. It

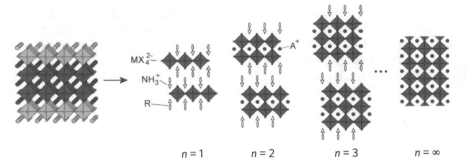

Fig. 4.2 Schematic demonstrating the relationship between the three-dimensional perovskites ($n = \infty$) and their layered counterparts ($n \geq 1$). Adapted from Ref. [47]

was shown that moving from a 3D to 2D perovskite structure widened the optical band gap considerably, with concomitant reduction in n-type conductivity. This trend was attributed, in part, to the presence of high exciton binding energies—often greater than 300 eV [54, 58]—which emerged due to spatial constraints and dielectric mismatch between the inorganic and organic components [59, 60]. Such properties are undesirable in photovoltaic absorbers as they result in a reduced V_{oc} that negatively impacts PCEs. Conversely, stability is greatest for the 2D $n = 1$ structure and decreases as the thickness of the perovskite layer is increased. Clearly, the primary challenge in designing novel layered perovskites is to optimise stoichiometry such that stability is balanced against optoelectronic performance.

In an attempt to offset the large band gap seen when adopting the 2D perovskite motif, the first layered perovskites employed in solar cells attempted to tune the electronic properties through shrewd choice of organic cation. Smith et al. synthesised an $n = 3$ perovskite containing the bulky $C_6H_5(CH_2)_2NH_3^+$ (PEA) cation, enabling a significant reduction in exciton binding energies (40 meV) but a band gap outside the ideal range (2.1 eV) [61]. Despite this, their cells performed reasonably well, recording efficiencies up to 4.73%, and displayed impressive stabilities, with little degradation seen after 40 days in air with 52% humidity. Similarly, $n = 3$ devices containing another large organic cation, $CH_3(CH_2)_3NH_3^+$, were synthesised by Cao et al. through a low-cost spin-coating process, achieving efficiencies of 4.02% [62].

Both the devices produced by Smith et al. and Cao et al. suffered from small charge-carrier diffusion lengths that severely impacted efficiencies [61, 62]. Recently, Quan et al. have produced the first hysteresis-free planar perovskite solar cell, based on a 2D absorbing layer containing PEA [49]. Density functional theory calculations were employed in tandem with complementary experimental studies on electronic and physical properties, to optimise the thickness of the perovskite layer to $40 < n < 60$. Van der Waals forces between the organic and perovskite layers were found to reduce the desorption rate of CH_3NH_3I (MAI)—the primary decomposition route of MAPI [63–65]—by six orders of magnitude. Furthermore, these quasi-2D devices successfully combined the ideal optoelectronic properties of MAPI with the stability of 2D perovskites, enabling an efficiency of 15.3% for the composition where $n = 60$ [49].

4.3   Tin-Based Perovskites

The instability of lead-based perovskite photovoltaics has raised concerns over the environmental effects if toxic lead were to leach into the environment [14]. Under EU law, commercial solar cells are exempt from the Restriction of Hazardous Substances Directive and are not regulated based on their lead content [66]. To ensure minimal effect on the environment, however, devices will require thorough encapsulation, which will increase the cost of manufacture. The last few years have therefore seen many attempts to produce lead-free perovskite photovoltaics. Tin is of particular interest due to its low cost and minimal toxicity [67]. Furthermore, it is isoelectric with lead and is therefore expected to show many of the same properties.

Unfortunately, the relatively facile oxidation of Sn^{2+} to Sn^{4+} has limited the adoption and development of tin hybrid perovskites [68, 69]. In 2015, Koh et al. demonstrated that by using small amounts of SnF_2 in the synthesis of formamidinium tin iodide ($HC(NH_2)_2SnI_3$, FASI) the reduction of Sn^{4+} to Sn^{2+} was promoted [70]. While this greater stability occurs at the expense of conductivity, this technique enabled the fabrication of FASI-based devices with efficiencies of 2.1%. FASI possesses a band gap of 1.4 eV—in the ideal range for a solar absorber—and, in contrast to its lead analogues, does not show other thermally accessible phases that might negatively impact real-world efficiencies [26, 29, 71]. Further work extended this method by incorporating pyrazine to prevent surface segregation of the SnF_2 [72]. This enabled improved power conversion efficiencies up to 4.8% (Fig. 4.3), with devices showing limited loss in performance when kept in ambient conditions for 100 days. More recently, FASI-based devices containing very small concentrations of a 2D tin-based perovskite have demonstrated efficiencies of 9.0% [73]. These

Fig. 4.3 J–V curves demonstrating the effect of pyrazine on FASnI$_3$ devices. Reprinted with permission from Ref. [72]. Copyright (2016) American Chemical Society

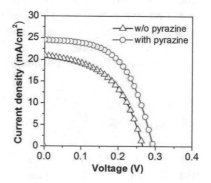

devices showed much greater stability in light and ambient conditions compared to films prepared in an analogous manner but instead using SnF_2.

The all-inorganic $CsSnI_3$ has also faced issues due to the undesired oxidation of Sn. Originally employed as a solid-state hole transporting layer for dye-sensitised solar cells [74], $CsSnI_3$ possesses a band gap of 1.30 eV and intrinsic p-type conductivity [75, 76]. While initial devices saw very poor performance due to the polycrystalline nature of the thin films [77], subsequent work employing SnF_2 as an additive enabled efficiencies up to 2.0% [78, 79]. Unfortunately, performance remains limited by poor open-circuit voltages of 0.24 V, about four times smaller than seen in MAPI-based devices [80, 81]. $CsSnI_3$-based devices face further challenges to due the formation of an alternate wide band gap secondary phase upon exposure to air or organic solvents [75, 82, 83]. Accordingly, $CsSnI_3$-based devices will require significant development if they are to become practically relevant.

When moving from $CH_3NH_3PbI_3$ to $CH_3NH_3SnI_3$, the band gap is reduced from 1.55 to 1.20 eV [68]. A similar trend is seen for the tin layered perovskites. For example, in the butylammonium (BA) layered perovskite series $(BA)_2(MA)_{n-1}Pb_nI_{3n+1}$, the band gaps for the compositions where $n = 1, 2$, and 3, are 2.43 eV, 2.17 eV, and 2.03 eV, respectively [62]. Moving to their tin analogues, the band gaps for the same n members were recently measured as 1.83 eV, 1.64 eV, and 1.50 eV, respectively [84]. Accordingly, the band gaps for the tin series are considerably closer to the ideal dictated by the Shockley–Quiesser limit.

In 2017, the compositions of $(BA)_2(MA)_{n-1}Sn_nI_{3n+1}$ where $n = 3$ and 4 were incorporated into photovoltaic devices containing mesoporous TiO_2 [84]. Using dimethyl sulfoxide (DMSO) as a solvent, the perovskite layer was grown such that the 2D sheets were vertically aligned perpendicular to the electrodes. This arrangement has been shown to be required for maximising the efficiency of layered devices. Despite this, the cells performed relatively poorly, achieving efficiencies up to 1.94%–2.53% for the $n = 3$ and 4 compounds, respectively. Promisingly, however, the stability of the devices was dramatically improved, with the cells retaining 90% of their peak performance after one month. In contrast, similarly prepared $CH_3NH_3SnI_3$ devices showed 0% efficiency after this time. Subsequent work on $(PEA)_2(FA)_8Sn_9I_{29}$ where the PEA cation is phenylethylammonium, enabled device efficiencies up to 5.94% [85]. Again, judicious choice of solvent was employed to ensure vertically aligned growth. Furthermore, the resulting thin films were shown to possess significantly improved stability compared to $FASnI_3$. While still very much in development, the progress seen in layered tin-based perovskites over the last few months provides hope for the future of lead-free perovskite photovoltaics.

4.4   Outlook

The success of the hybrid perovskites is evident. That a little-known and understudied material can achieve power conversion efficiencies over 22% in less than ten years of development is remarkable. Despite this, there remain several key issues that must be addressed if perovskite-based photovoltaics are to be successfully commercialised. Crucially, their long-term stability presents a major roadblock, which, even after significant research attention, still plagues the perovskite community. The recent emergence of the layered perovskites provides a glimmer of hope. In particular, "quasi-layered" materials combine the dramatic increase in stability of the purely layered structures with the optimal optoelectronic properties possessed by the three-dimensional perovskites. Accordingly, further work tuning layer thickness may prove incredibly fruitful.

While currently the use of lead in photovoltaic devices is not regulated, this may not remain the case in the future. Despite the isoelectronic nature of the group 14 metals, tin-based perovskite absorbers cannot yet match the performance of their lead counterparts. Here, progress has been hampered by the sensitivity of Sn^{2+} to oxidation, which complicates device fabrication and stability. Recent months have seen significant improvements in this area, particularly for FASI-based devices, which have now achieved efficiencies of 9.0%. Regardless, it is essential that alternative schemes for improving the stability of these materials be developed.

References

1. Kim H-S, Park N-G (2014) Parameters affecting I-V hysteresis of $CH_3NH_3PbI_3$ perovskite solar cells: effects of perovskite crystal size and mesoporous TiO_2 layer. J Phys Chem Lett 5:2927–2934
2. Beilsten-Edmands J, Eperon GE, Johnson RD, Snaith HJ, Radaelli PG (2015) Non-ferroelectric nature of the conductance hysteresis in $CH_3NH_3PbI_3$ perovskite-based photovoltaic devices. Appl Phys Lett 106:173502
3. Frost JM, Walsh A (2016) What is moving in hybrid halide perovskite solar cells? Acc Chem Res 49:528–535
4. Azpiroz JM, Mosconi E, Bisquert J, De Angelis F (2015) Defect migration in methylammonium lead iodide and its role in perovskite solar cell operation. Energy Environ Sci 8:2118–2127
5. Eames C, Frost JM, Barnes PR, O'regan BC, Walsh A, Islam MS (2015) Ionic transport in hybrid lead iodide perovskite solar cells. Nat Commun 6:7497
6. Tidhar Y, Edri E, Weissman H, Zohar D, Hodes G, Cahen D, Rybtchinski B, Kirmayer S (2014) Crystallization of methyl ammonium lead halide perovskites: implications for photovoltaic applications. J Am Chem Soc 136:13249–13256
7. Jeon NJ, Noh JH, Kim YC, Yang WS, Ryu S, Seok SI (2014) Solvent engineering for high-performance inorganic-organic hybrid perovskite solar cells. Nat Mater 13:897–903
8. Leo K (2015) Perovskite photovoltaics: signs of stability. Nat Nanotechnol 10:574–575
9. Hailegnaw B, Kirmayer S, Edri E, Hodes G, Cahen D (2015) Rain on methyl-ammonium-lead-iodide based perovskites: possible environmental effects of perovskite solar cells. J Phys Chem Lett 6:1543–1547
10. Bass KK, McAnally RE, Zhou S, Djurovich PI, Thompson ME, Melot BC (2014) Influence of moisture on the preparation, crystal structure, and photophysical properties of organohalide perovskites. Chem Commun 50:15819–15822
11. Mosconi E, Azpiroz JM, De Angelis F (2015) Ab initio molecular dynamics simulations of methylammonium lead iodide perovskite degradation by water. Chem Mater 27:4885–4892
12. Christians JA, Manser JS, Kamat PV (2015) Multifaceted excited state of $CH_3NH_3PbI_3$. charge separation, recombination, and trapping. J Phys Chem Lett 6:2086–2095

13. Bryant D, Aristidou N, Pont S, Sanchez-Molina I, Chotchunangatchaval T, Wheeler S, Durrant JR, Haque SA (2016) Light and oxygen induced degradation limits the operational stability of methylammonium lead triiodide perovskite solar cells. Energy Environ Sci 9:1655–1660
14. Conings B, Drijkoningen J, Gauquelin N, Babayigit A, D'Haen J, D'Olieslaeger L, Ethirajan A, Verbeeck J, Manca J, Mosconi E, Angelis FD, Boyen H-G (2015) Intrinsic thermal instability of methylammonium lead trihalide perovskite. Adv Energy Mater 5:1500477
15. Juarez-Perez EJ, Hawash Z, Raga SR, Ono LK, Qi Y (2016) Thermal degradation of $CH_3NH_3PbI_3$ perovskite into NH_3 and CH_3I gases observed by coupled thermogravimetry-mass spectrometry analysis. Energy Environ Sci 9:3406–3410
16. Niu G, Li W, Meng F, Wang L, Dong H, Qiu Y (2014) Study on the stability of $CH_3NH_3PbI_3$ films and the effect of post-modification by aluminum oxide in all-solid-state hybrid solar cells. J Mater Chem A 2:705–710
17. Pisoni A, Jaćimović J, Barišić OS, Spina M, Gaál R, Forró L, Horváth E (2014) Ultra-low thermal conductivity in organic-inorganic hybrid perovskite $CH_3NH_3PbI_3$. J Phys Chem Lett 5:2488–2492
18. Zhang Y-Y, Chen S, Xu P, Xiang H, Gong X-G, Walsh A, Wei S-H (2018) Intrinsic instability of the hybrid halide perovskite semiconductor $CH_3NH_3PbI_3$. Chin Phys Lett 35:036104
19. Ganose AM, Savory CN, Scanlon DO (2015) $(CH_3NH_3)_2Pb(SCN)_2I_2$: A more stable structural motif for hybrid halide photovoltaics? J Phys Chem Lett 6:4594–4598
20. Habisreutinger SN, McMeekin DP, Snaith HJ, Nicholas RJ (2016) Research update: strategies for improving the stability of perovskite solar cells. APL Mater 4:091503
21. Mitzi DB (2001) Templating and structural engineering in organic-inorganic perovskites. J Chem Soc Dalton Trans 1–12
22. Boix PP, Agarwala S, Koh TM, Mathews N, Mhaisalkar SG (2015) Perovskite solar cells: beyond methylammonium lead iodide. J Phys Chem Lett 6:898–907
23. Saparov B, Mitzi DB (2016) Organic-inorganic perovskites: structural versatility for functional materials design. Chem Rev 116:4558–4596
24. Amat A, Mosconi E, Ronca E, Quarti C, Umari P, Nazeeruddin MK, Grätzel M, De Angelis F (2014) Cation-induced band-gap tuning in organohalide perovskites: Interplay of spin-orbit coupling and octahedra tilting. Nano Lett 14:3608–3616
25. Eperon GE, Stranks SD, Menelaou C, Johnston MB, Herz LM, Snaith HJ (2014) Formamidinium lead trihalide: a broadly tunable perovskite for efficient planar heterojunction solar cells. Energy Environ Sci 7:982–988
26. Stoumpos CC, Malliakas CD, Kanatzidis MG (2013) Semiconducting tin and lead iodide perovskites with organic cations: Phase transitions, high mobilities, and near-infrared photoluminescent properties. Inorg Chem 52:9019–9038
27. Weller MT, Weber OJ, Frost JM, Walsh A (2015) Cubic perovskite structure of black formamidinium lead iodide, α-[$HC(NH_2)_2$]PbI_3, at 298 K. J Phys Chem Lett 6:3209–3212
28. Binek A, Hanusch FC, Docampo P, Bein T (2015) Stabilization of the trigonal high-temperature phase of formamidinium lead iodide. J Phys Chem Lett 6:1249–1253
29. Pellet N, Gao P, Gregori G, Yang T-Y, Nazeeruddin MK, Maier J, Grätzel M (2014) Mixed-organic-cation perovskite photovoltaics for enhanced solar-light harvesting. Angew Chem Int Ed 53:3151–3157
30. Brgoch J, Lehner A, Chabinyc M, Seshadri R (2014) Ab initio calculations of band gaps and absolute band positions of polymorphs of $RbPbI_3$ and $CsPbI_3$: implications for main-group halide perovskite photovoltaics. J Phys Chem C 118:27721–27727
31. Sutton RJ, Eperon GE, Miranda L, Parrott ES, Kamino BA, Patel JB, Hörantner MT, Johnston MB, Haghighirad AA, Moore DT, Snaith HJ (2016) Bandgap-tunable cesium lead halide perovskites with high thermal stability for efficient solar cells. Adv Energy Mater 6:1502458
32. Eperon GE, Paterno GM, Sutton RJ, Zampetti A, Haghighirad AA, Cacialli F, Snaith HJ (2015) Inorganic caesium lead iodide perovskite solar cells. J Mater Chem A 3:19688–19695
33. Protesescu L, Yakunin S, Bodnarchuk MI, Krieg F, Caputo R, Hendon CH, Yang RX, Walsh A, Kovalenko MV (2015) Nanocrystals of cesium lead halide perovskites ($CsPbX_3$, X= Cl, Br, and I): novel optoelectronic materials showing bright emission with wide color gamut. Nano Lett 15:3692–3696

34. Butler KT, Frost JM, Walsh A (2015) Band alignment of the hybrid halide perovskites $CH_3NH_3PbCl_3$, $CH_3NH_3PbBr_3$ and $CH_3NH_3PbI_3$. Mater Horiz 2:228–231
35. Stranks SD, Eperon GE, Grancini G, Menelaou C, Alcocer MJP, Leijtens T, Herz LM, Petrozza A, Snaith HJ (2013) Electron-hole diffusion lengths exceeding 1 micrometer in an organometal trihalide perovskite absorber. Science 342:341–344
36. Yantara N, Fang Y, Chen S, Dewi HA, Boix PP, Mhaisalkar SG, Mathews N (2015) Unravelling the effects of Cl addition in single step $CH_3NH_3PbI_3$ perovskite solar cells. Chem Mater 27:2309–2314
37. Noh JH, Im SH, Heo JH, Mandal TN, Seok SI (2013) Chemical management for colorful, efficient, and stable inorganic-organic hybrid nanostructured solar cells. Nano Lett 13:1764–1769
38. Rehman W, Milot RL, Eperon GE, Wehrenfennig C, Boland JL, Snaith HJ, Johnston MB, Herz LM (2015) Charge-carrier dynamics and mobilities in formamidinium lead mixed-halide perovskites. Adv Mater 27:7938–7944
39. Brivio F, Caetano C, Walsh A (2016) Thermodynamic origin of photoinstability in the $CH_3NH_3Pb(I_{1-x}Br_x)_3$ hybrid halide Perovskite Alloy. J Phys Chem Lett 7:1083–1087
40. McMeekin DP, Sadoughi G, Rehman W, Eperon GE, Saliba M, Hörantner MT, Haghighirad A, Sakai N, Korte L, Rech B, Johnston MB, Herz LM, Snaith HJ (2016) A mixed-cation lead mixed-halide perovskite absorber for tandem solar cells. Science 351:151–155
41. Duong T et al (2017) Rubidium multiplication perovskite with optimized bandgap for perovskite-silicon tandem with over 26% efficiency. Adv Energy Mater 7:1700228
42. Koh TM, Thirumal K, Soo HS, Mathews N (2016) Multidimensional perovskites: a mixed cation approach towards ambient stable and tunable perovskite photovoltaics. ChemSusChem 9:2541–2558
43. Ganose AM, Savory CN, Scanlon DO (2017) Beyond methylammonium lead iodide: prospects for the emergent field of ns^2 containing solar absorbers. Chem Commun 53:20–44
44. Gregor K, Shijing S, Anthony KC (2014) Solid-state principles applied to organic-inorganic perovskites: new tricks for an old dog. Chem Sci 5:4712–4715
45. Kieslich G, Sun S, Cheetham T (2015) An extended tolerance factor approach for organic-inorganic perovskites. Chem Sci 6:3430–3433
46. Mitzi DB, Dimitrakopoulos CD, Kosbar LL (2001) Structurally tailored organic-inorganic perovskites: optical properties and solution-processed channel materials for thin-film transistors. Chem Mater 13:3728–3740
47. Mitzi DB, Chondroudis K, Kagan CR (2001) Organic-inorganic electronics. IBM J Res Dev 45:29–45
48. Mitzi DB, Medeiros DR, Malenfant PRL (2002) Intercalated organic-inorganic perovskites stabilized by fluoroaryl–aryl interactions. Inorg Chem 41:2134–2145
49. Quan LN, Yuan M, Comin R, Voznyy O, Beauregard EM, Hoogland S, Buin A, Kirmani AR, Zhao K, Amassian A, Kim DH, Sargent EH (2016) Ligand-stabilized reduced-dimensionality perovskites. J Am Chem Soc 138:2649–2655
50. Jiang W, Ying J, Zhou W, Shen K, Liu X, Gao X, Guo F, Gao Y, Yang T (2016) A new layered nano hybrid perovskite film with enhanced resistance to moisture-induced degradation. Chem Phys Lett 658:71–75
51. Mitzi DB, Feild CA, Harrison WTA, Guloy AM (1994) Conducting tin halides with a layered organic-based perovskite structure. Nature 369:467–469
52. Mitzi DB (1999) In: Karlin KD (ed) Progress in inorganic chemistry, vol 48. Wiley, pp 1–121 (Chapter 1)
53. Wang S, Mitzi DB, Feild CA, Guloys A (1995) Synthesis and characterization of $[NH_2C(I)=NH_2]_3MI_5$ (M = Sn, Pb): stereochemical activity in divalent tin and lead halides containing single (1 10) perovskite sheets. J Am Chem Soc 117:5297–5302
54. Mitzi DB (1996) Synthesis, crystal structure, and optical and thermal properties of $(C_4H_9NH_3)_2MI_4$ (M = Ge, Sn, Pb). Chem Mater 8:791–800
55. Chondroudis K, Mitzi DB (1999) Electroluminescence from an organic-inorganic perovskite incorporating a quaterthiophene dye within lead halide perovskite layers. Chem Mater 11:3028–3030

56. Mitzi DB (2000) Organic-inorganic perovskites containing trivalent metal halide layers: the templating influence of the organic cation layer. Inorg Chem 39:6107–6113
57. Mitzi DB (2001) Thin-film deposition of organic-inorganic hybrid materials. Chem Mater 13:3283–3298
58. Kitazawa N, Watanabe Y (2010) Optical properties of natural quantum-well compounds (C_6H_5-C_nH_{2n}-NH_3)$_2$PbBr$_4$ (n = 1–4). J Phys Chem Solids 71:797–802
59. Ishihara T, Takahashi J, Goto T (1990) Optical properties due to electronic transitions in two-dimensional semiconductors ($C_nH_{2n+1}NH_3$)$_2$PbI$_4$. Phys Rev B 42:11099–11107
60. Muljarov EA, Tikhodeev SG, Gippius NA, Ishihara T (1995) Excitons in self-organized semiconductor/insulator superlattices: PbI-based perovskite compounds. Phys Rev B 51:14370–14378
61. Smith IC, Hoke ET, Solis-Ibarra D, McGehee MD, Karunadasa HI (2014) A layered hybrid perovskite solar-cell absorber with enhanced moisture stability. Angew Chem Int Ed 53:11232–11235
62. Cao DH, Stoumpos CC, Farha OK, Hupp JT, Kanatzidis MG (2015) 2D homologous perovskites as light-absorbing materials for solar cell applications. J Am Chem Soc 137:7843–7850
63. Yang Z, Chueh C-C, Zuo F, Kim JH, Liang P-W, Jen AK-Y (2015) High-performance fully printable perovskite solar cells via blade-coating technique under the ambient Condition. Adv Energy Mater 5:1500328
64. Dualeh A, Gao P, Seok SI, Nazeeruddin MK, Grätzel M (2014) Thermal behavior of methylammonium lead-trihalide perovskite photovoltaic light harvesters. Chem Mater 26:6160–6164
65. Christians JA, Herrera PAM, Kamat PV (2015) Transformation of the excited state and photovoltaic efficiency cells $_3NH_3PbI_3$ perovskite upon controlled exposure to humidified air. J Am Chem Soc 137:1530–1538
66. Directive 2011/65/EU of the European Parliament and of the Council of 8 June 2011 on the restriction of the use of certain hazardous substances in electrical and electronic equipment (recast). http://eur-lex.europa.eu/legal-content/en/TXT/?uri=celex:32011L0065 (2011)
67. Winship K (1987) Toxicity of tin and its compounds. Adverse drug react. Acute Poisoning Rev 7:19–38
68. Umari P, Mosconi E, De Angelis F (2014) Relativistic GW calculations on $CH_3NH_3PbI_3$ and $CH_3NH_3SnI_3$ perovskites for solar cell applications. Sci Rep 4:4467
69. Brenner TM, Egger DA, Kronik L, Hodes G, Cahen D (2016) Hybrid organic-inorganic perovskites: low-cost semiconductors with intriguing charge-transport properties. Nat Rev Mater 1:15007
70. Koh TM, Krishnamoorthy T, Yantara N, Shi C, Leong WL, Boix PP, Grimsdale AC, Mhaisalkar SG, Mathews N (2015) Formamidinium tin-based perovskite with low E_g for photovoltaic applications. J Mater Chem A 3:14996–15000
71. Koh TM, Fu K, Fang Y, Chen S, Sum T, Mathews N, Mhaisalkar SG, Boix PP, Baikie T (2013) Formamidinium-containing metal-halide: an alternative material for near-IR absorption perovskite solar cells. J Mater Chem C 118:16458–16462
72. Lee SJ, Shin SS, Kim YC, Kim D, Ahn TK, Noh JH, Seo J, Seok SI (2016) Fabrication of efficient formamidinium tin iodide perovskite solar cells through SnF$_2$-pyrazine complex. J Am Chem Soc 138:3974–3977
73. Shao S, Liu J, Portale G, Fang H-H, Blake GR, ten Brink GH, Koster LJA, Loi MA (2018) Highly reproducible Sn-based hybrid perovskite solar cells with 9% efficiency. Adv Energy Mater 8:1702019
74. Chung I, Lee B, He J, Chang RP, Kanatzidis MG (2012) All-solid-state dye-sensitized solar cells with high efficiency. Nature 485:486–489
75. Chung I, Song JH, Im J, Androulakis J, Malliakas CD, Li H, Freeman AJ, Kenney JT, Kanatzidis MG (2012) CsSnI$_3$: semiconductor or metal? High electrical conductivity and strong near-infrared photoluminescence from a single material. High hole mobility and phase-transitions. J Am Chem Soc 134:8579–8587
76. Zhou Y, Garces HF, Senturk BS, Ortiz AL, Padture NP (2013) Room temperature "one-pot" solution synthesis of nanoscale CsSnI$_3$ orthorhombic perovskite thin films and particles. Mater Lett 110:127–129

77. Chen Z, Wang JJ, Ren Y, Yu C, Shum K (2012) Schottky solar cells based on $CsSnI_3$ thin-films. Appl Phys Lett 101:093901
78. Sabba D, Mulmudi HK, Prabhakar RR, Krishnamoorthy T, Baikie T, Boix PP, Mhaisalkar S, Mathews N (2015) Impact of anionic Br^- substitution on open circuit voltage in lead free perovskite $(CsSnI_{3-x}Br_x)$ solar cells. J Phys Chem C 119:1763–1767
79. Kumar MH, Dharani S, Leong WL, Boix PP, Prabhakar RR, Baikie T, Shi C, Ding H, Ramesh R, Asta M, Graetzel M, Mhaisalkar SG, Mathews N (2014) Lead-free halide perovskite solar cells with high photocurrents realized through vacancy modulation. Adv Mater 26:7122–7127
80. Dharani S, Mulmudi HK, Yantara N, Trang PTT, Park NG, Graetzel M, Mhaisalkar S, Mathews N, Boix PP (2014) High efficiency electrospun TiO_2 nanofiber based hybrid organic-inorganic perovskite solar cell. Nanoscale 6:1675–1679
81. Kumar MH, Yantara N, Dharani S, Graetzel M, Mhaisalkar S, Boix PP, Mathews N (2013) Flexible, low-temperature, solution processed ZnO-based perovskite solid state solar cells. Chem Commun 49:11089–11091
82. Yu C, Ren Y, Chen Z, Shum K (2013) First-principles study of structural phase transitions in $CsSnI_3$. J Appl Phys 114:163505
83. da Silva EL, Skelton JM, Parker SC, Walsh A, Silva EL, Skelton JM, Parker SC, Walsh A (2015) Phase stability and transformations in the halide perovskite $CsSnI_3$. Phys Rev B 91:1–12
84. Cao DH, Stoumpos CC, Yokoyama T, Logsdon JL, Song T-B, Farha OK, Wasielewski MR, Hupp JT, Kanatzidis MG (2017) Thin films and solar cells based on semiconducting two-dimensional Ruddlesden-Popper $(CH_3(CH_2)_3NH_3)_2(CH_3NH_3)_{n-1}Sn_nI_{3n+1}$ perovskites. ACS Energy Lett 2:982–990
85. Liao Y, Liu H, Zhou W, Yang D, Shang Y, Shi Z, Li B, Jiang X, Zhang L, Quan LN, Quintero-Bermudez R, Sutherland BR, Mi Q, Sargent EH, Ning Z (2017) Highly oriented low-dimensional tin halide perovskites with enhanced stability and photovoltaic performance. J Am Chem Soc 139:6693–6699

Chapter 5
Pseudohalide Perovskite Absorbers

5.1 Introduction

The substitution of iodine for thiocyanate (SCN) has recently emerged as a novel route to increasing the stability of MAPI-based devices. References [2–6] In 2015, Chen et al. showed that inclusion of SCN^-, a pseudohalide with an ionic radius similar to that of I- (217 pm vs. 220 pm) [7], to form $CH_3NH_3PbI_{3-x}SCN_x$ promoted larger crystal domains with fewer trap sites than in undoped MAPI samples [3]. Their devices achieved efficiencies up to 11 % at an optimum level of 5 % SCN incorporation and possessed enhanced stability, greater reproducibility, and reduced levels of hysteresis than comparable MAPI films [8, 9]. The effect of SCN as a *dopant* in MAPI was investigated by Halder et al., who observed a ten-fold enhancement in the intensity of the photoluminescence response, coupled with a band gap widening of 8 meV [4]. Jiang et al. provided the first structural characterisation of pseudohalide containing perovskites, suggesting that $CH_3NH_3Pb(SCN)_2I$ films crystallised in the cubic perovskite structure [5]. Their films, with a band gap of 1.53 eV, were significantly more stable in air with 95% humidity than pure MAPI, the reason for which was not elucidated. [10] Devices containing $CH_3NH_3Pb(SCN)_2I$ films showed an efficiency of 8.3 %, with a greater V_{oc} (0.87 eV vs. 0.80 eV) but smaller fill factor (FF, 52 vs. 63) than similarly prepared MAPI films [5]. Unfortunately, this was the first report of SCN incorporation on the X site in the cubic perovskite structure, leading to doubts of the structural characterisation within the hybrid-perovskite community.

Subsequently, Daub and Hillebrecht showed that the reaction of $Pb(SCN)_2$ and CH_3NH_3I results in the formation of $(CH_3NH_3)_2Pb(SCN)_2I_2$, which we will denote MAPSI for simplicity [6]. MAPSI crystallises in a layered orthorhombic structure ($Pnm2_1$), similar to the 2D $n = 1$ perovskites, with Pb octahedrally coordinated to

---

Parts of this chapter have been reproduced with permission from [1] (published by The Royal Society of Chemistry) and [2] (published by the American Chemical Society).

A. Ganose, *Atomic-Scale Insights into Emergent Photovoltaic Absorbers*, Springer Theses, https://doi.org/10.1007/978-3-030-55708-9_5

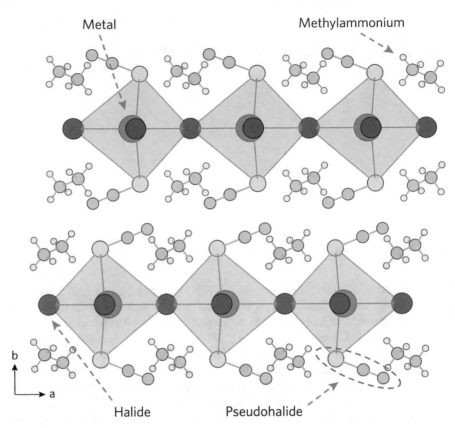

Fig. 5.1 Crystal structure of $(CH_3NH_3)_2Pb(SCN)_2I_2$ viewed along the [001] direction. Pb, I, S, C, N, and H atoms denoted by dark grey, purple, yellow, red, blue, and pink spheres, respectively. The octahedral nature of Pb is illustrated using green polyhedra

two axial SCN and four equatorial I, with the methylammonium sandwiched between the layers (Fig. 5.1). Crucially, the authors demonstrated that the X-ray diffraction (XRD) pattern for MAPSI matched the XRD pattern of the $CH_3NH_3Pb(SCN)_2I$ films produced by Jiang et al. [5].

There has been some debate as to the magnitude of MAPSI's optical band gap: initial work into thiocyanate incorporation indicated a direct optical gap of 1.57 eV [6, 11], however, a study performed by Xiao et al. reported thin films with indirect and direct gaps of 2.04 eV and 2.11 eV, respectively [12]. Recently, Umeyama et al. have suggested MAPSI possesses a larger band gap than originally thought, noticing a red-to-black piezochromic response upon compression (2.6 GPa) [13]. While the cause of this inconsistency is currently unknown, there have been suggestions that contamination of MAPSI with slight amounts of MAPI may play a role. An alternative suggestion, proposed by Younts et al., relies on the observation of highly efficient triplet state formation in MAPSI thin films [14]. They report a triplet energy of 1.64 eV, with phosphorescence over 47 times more intense than the band gap fluorescence, thereby explaining some of the previous photoluminescence results that have frustrated band gap determination. Despite this, a band gap of ~2.1 eV and exciton binding energies less than 200 meV [13] are uncommon in two-dimensional perovskites, which often show band gaps >2.7 meV with 350 meV exciton binding energies [15]. As such, MAPSI presents an intriguing material that deserves further study.

5.2 Methodology

Calculations were performed using the Vienna Ab initio Simulation Package. A **k**-point mesh of Γ-centred $1 \times 4 \times 4$ and plane wave cutoff of 400 eV was found to converge the 50 atom unit cell of $(CH_3NH_3)_2Pb(SCN)_2I_2$ and all analogues to within 1 meV/atom. During geometry optimisations, the cutoff was increased to 520 eV to avoid errors resulting from Pulay stress [16]. The structures were deemed converged when the forces totalled <10 meV $\overset{\circ}{A}^{-1}$.

Several functionals were used in this work: For geometry relaxations, PBEsol and PBE were employed, with and without the addition of Grimme's D3 dispersion correction. Electronic properties were calculated using HSE06 with the addition of spin–orbit coupling effects (HSE06+SOC). The Brillouin zone for the $Pnm2_1$ space group, indicating the high-symmetry points explored in the band structure, is provided in Fig. 5.2. Density functional perturbation theory (DFPT) was employed, in combination with the PBEsol functional, to calculate the ionic contribution to the dielectric constants, with a denser **k**-point mesh of Γ-centred $3 \times 6 \times 6$ required to achieve convergence.

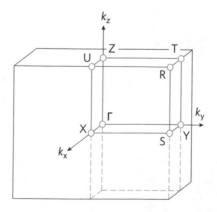

Fig. 5.2 Brillouin zone of the $Pnm2_1$ space group. Coordinates of the high symmetry points used for the band structures and effective masses: $\Gamma = (0, 0, 0)$; $Y = (1/2, 0, 0)$; $X = (0, 1/2, 0)$; $Z = (0, 0, 1/2)$; $U = (0, 1/2, 1/2)$; $T = (1/2, 0, 1/2)$; $S = (1/2, 1/2, 0)$; $R = (0, 1/2, 1/2)$

For band alignment calculations, the core-level alignment approach of Wei and Zunger was employed [17], using a slab model with 30 Å of vacuum and a 25 Å thick slab. The slab was cleaved along the non-polar (010) surface, due to the absence of any dangling bonds. Due to the size of the model, which precluded the use of HSE06+SOC, band alignment calculations were performed using HSE06, with an explicit correction to the band gap and valence band maximum position taken from the HSE06+SOC calculated bulk.

Defect calculations were performed in a $1 \times 3 \times 3$ supercell containing 450 atoms, using the PBEsol functional and a Γ-centred $2 \times 2 \times 2$ **k**-point mesh. The defect energies were corrected to account for use of a finite-sized supercell, using the potential level alignment, band-filling and image-charge corrections described in Chap. 3.

5.3 Results

5.3.1 Geometric Structure

The crystal structure of MAPSI was minimised using three functionals, PBEsol, PBEsol with the addition of Grimme's D3 dispersion correction (PBEsol+D3) and PBE with the D3 correction (PBE+D3), with the results given in Table 5.1. Both PBEsol and PBE+D3 show good agreement with the experimental single crystal structure [6], with PBEsol+D3 severely underestimating the lattice constants. The a lattice parameter is underestimated across every functional, suggesting that thermal effects may play a role in determining the interlayer separation. The PBEsol relaxed structure possessed lattice constants closest to experiment and was therefore used for all subsequent calculations.

5.3.2 Electronic Properties

Calculations on the electronic structure of MAPSI were performed using the HSE06 hybrid DFT functional, with explicit treatment of spin–orbit coupling (SOC), to

Table 5.1 Lattice parameters of $(CH_3NH_3)_2Pb(SCN)_2I_2$ calculated using PBEsol, PBEsol+D3 and PBE+D3. Experimental error or percentage difference from experiment in parentheses. All cell angles were found to be 90°

	a (Å)	b (Å)	c (Å)
PBEsol	6.230 (−0.59%)	18.268 (−1.68%)	6.475 (+0.14%)
PBEsol+D3	6.134 (−2.12%)	17.657 (−4.97%)	6.388 (−1.21%)
PBE+D3	6.274 (+0.11%)	18.232 (−1.87%)	6.525 (+0.91%)
Experiment [6]	6.267(7)	18.580(2)	6.466(6)

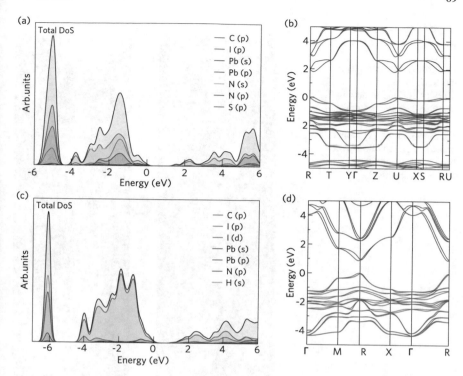

Fig. 5.3 Comparison of the electronic structures of $(CH_3NH_3)_2Pb(SCN)_2I_2$ (MAPSI) and $CH_3NH_3PbI_3$ (MAPI) calculated using HSE06 with the addition of spin–orbit coupling. **a, c** Projected density of states and **b, d** band structure for MAPSI and MAPI, respectively. The VBM is set to 0 eV in all cases. The conduction and valence bands of **b** and **d** are denoted by orange and blue lines, respectively

ensure the accurate treatment of relativistic effects known to strongly affect the electronic structure of the lead iodide perovskites. The band gap of MAPSI was found to be 1.79 eV, significantly smaller than the reported experimental optical band gap of ~2.1 eV [13, 14]. This underestimation of the band gap is similar to that seen in MAPI, where tuning of the amount of Hartree–Fock exchange, α, is needed to adjust the calculated band gap to the experimental gap. In this work we have chosen not to artificially fit the band gap parameter as this discrepancy may arise from several factors, including the underestimation of the a lattice parameter.

The projected density of states calculated by HSE06+SOC is shown in Fig. 5.3a. The valence band maximum (VBM) is dominated by I p states, with some contribution from Pb s states. Similar to electronic structure of MAPI, the conduction band minimum (CBM) is composed almost entirely of Pb p (Fig. 5.3c). Comparing the electronic structures of MAPSI and MAPI, the main difference is the introduction of N p and S p states around 1 eV below the VBM.

Fig. 5.4 Charge density isosurfaces of the **a** VBM and **b** CBM of $(CH_3NH_3)_2Pb(SCN)_2I_2$, calculated using HSE06+SOC. Low and high electron density are indicated by blue and red, respectively. Pb, S, N, C, I, and H atoms shown by grey, yellow, blue, brown, purple and pink spheres, respectively

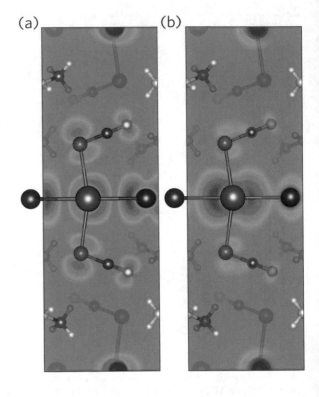

In semiconductors, the absolute ionisation potential—and therefore the position of the VBM—has been shown to depend on the amount of d–p repulsion present in the system [18]. Comparing the Pb d states of MAPSI and MAPI reveals an average shift in energy of 0.15 eV. This slight increase in p–d repulsion is insufficient to account for the smaller band gap of MAPSI compared to other 2D hybrid perovskites. Instead, the presence of the SCN states at the top of the valence band must act to push up the VBM, allowing for a small band gap in spite of its layered nature. Analysis of the S–C–N bond lengths corroborates this picture, revealing an increase in the covalent character of the pseudohalide, as indicated by the lengthening of the C–N (1.17 Å) and shortening of the S–C bonds (1.60 Å) in MAPSI compared to the ionic AgSCN, where the bond lengths are 1.14 Å and 1.78 Å, respectively [19]. The thiocyanate is therefore actively involved in bonding in the MAPSI structure, as evidenced by the charge density isosurfaces of the CBM and VBM (Fig. 5.4), and in contrast to other polyanion containing MAPI-based systems [20]. From the isosurfaces, the contributions to the VBM from the S $3p$ and N $2p$ orbitals can be seen clearly, whereas the CBM is dominated by Pb $6p$ states.

The band structure of MAPSI, calculated using HSE06+SOC, is presented in Fig. 5.3b. The fundamental band gap is 1.79 eV, with the VBM and CBM situated just off the R $(\frac{\bar{1}}{2}, \frac{1}{2}, \frac{1}{2})$ and U $(0, \frac{1}{2}, \frac{1}{2})$ points, respectively. Very little dispersion is seen across the X–S and R–U directions, as expected as these paths cross the layers in the [010] direction. Reasonably strong Rashba splitting is seen in the conduction band and, to a lesser extent, in the valence band, due to the lack of inversion symmetry in the MAPSI crystal structure. As previously discussed, such Rashba splitting has been shown to result in dramatically reduced rates of radiative recombination and longer charge-carrier diffusion lengths in MAPI [21].

The effective masses of the CBM and VBM were found to be small — $0.14 m_0$ and $0.20 m_0$, respectively. These are similar in magnitude to those of MAPI ($0.15 m_0$ and $0.12 m_0$, for electrons and holes, respectively) [22] and indicate that both electrons and holes should be mobile in the system. However, we stress that due to the 2D crystal structure and limited dispersion seen in the [010] direction, conductivity is expected to be strongly anisotropic. A large dielectric constant, ε_r, is increasingly considered a desirable property for a solar absorber due to it's role in screening charged defects and aiding electron–hole separation. The dielectric response of MAPSI was found to be reasonably anisotropic, with values of 16.9, 7.4, and 16.7 calculated for the [100], [010], and [001] directions, respectively. As expected due to the loss in connectivity of the iodide octahedra, these are smaller than those found in the cubic hybrid perovskites, which are often 60–70 [22], but are greater than in other third-generation absorbers such as CZTS ($\sim$9) [23].

Spin–orbit coupling was found to play a significant role in the electronic structure, with a 0.68 eV relativistic lowering of the conduction band observed (depicted in Fig. A.4 of Appendix A). This demonstrates that inclusion of relativistic effects is vital to accurately describe the electronic structure of MAPSI. We note that while many-body effects, such as electron–hole interactions, will also likely play a fundamental role in this system [24, 25], the size of the MAPSI unit cell (50 atoms) precludes their inclusion in this study.

5.3.3 Stability

MAPI's chemical stability has been the topic of much debate over the past five years [26, 27]. While it is known that atmospheric moisture content causes rapid degradation of MAPI films [10, 28], there is growing theoretical and experimental evidence that suggests the intrinsic material is itself thermodynamically unstable with respect to phase separation into $CH_3NH_3I_3$ and PbI_2 (the synthetic starting materials) [27, 29, 30]. To investigate the stability of MAPSI, we have trialled three decomposition pathways, in addition to the decomposition route of MAPI as a comparison:

$$CH_3NH_3PbI_3 \rightarrow CH_3NH_3I + PbI_2,$$
$$\Delta_d H = -0.09 \, eV \tag{5.1}$$
$$(CH_3NH_3)_2Pb(SCN)_2I_2 \rightarrow 2\,CH_3NH_3I + Pb(SCN)_2,$$
$$\Delta_d H = 0.38 \, eV \tag{5.2}$$
$$(CH_3NH_3)_2Pb(SCN)_2I_2 \rightarrow 2\,CH_3NH_3(SCN) + PbI_2,$$
$$\Delta_d H = 1.97 \, eV \tag{5.3}$$

$$(CH_3NH_3)_2Pb(SCN)_2I_2 \rightarrow CH_3NH_3SCN + \tfrac{1}{2}CH_3NH_3PbI_3 \qquad (5.4)$$
$$\tfrac{1}{2}CH_3NH_3I + \tfrac{1}{2}Pb(SCN)_2,$$
$$s\Delta_d H = 1.20\,eV$$

Unlike in MAPI, where the negative enthalpy of decomposition ($\Delta_d H$) indicates it is energetically favourable to spontaneously decompose into $CH_3NH_3I_3$ and PbI_2, the decomposition routes for MAPSI are all positive, revealing that phase separation is unfavourable. These results suggest a likely source of the increase in stability reported for hybrid halide materials with SCN incorporation.

5.3.4   Intrinsic Defects

The fundamental defect chemistry of a material will determine its ability to show intrinsic n- or p-type conductivity. Accordingly, understanding the defect behaviour of photovoltaics is instrumental in assessing their real-world performance. To this end, we have investigated a range of intrinsic donor and acceptor vacancy defects in MAPSI, namely: two iodine vacancies (due to the presence of two symmetrically inequivalent iodine sites in the MAPSI structure), V_I^1 and V_I^2; a lead vacancy, V_{Pb}; a methylammonium vacancy, V_{MA}; and an SCN vacancy, V_{SCN}. The PBEsol calculated charge-state transition level diagram is presented in Fig. 5.5.

Of the two V_I defects, one has both transition levels resonant in the conduction band, the other possesses a resonant $-1/0$ level and a reasonably shallow $0/+1$ transition state 0.15 eV below the CBM. Additionally, the V_{SCN} donor defect possesses one $0/+1$ transition level resonant in the conduction band. Considering the acceptor defects: the methylammonium vacancy, V_{MA}, is resonant in valence band, as expected due to the limited bonding interaction between the organic and inorganic components. In contrast, V_{Pb} possesses a shallow $-1/0$ transition level 0.11 eV above the valence band maximum but an ultra-deep $-2/-1$ level 0.68 eV below the conduction band edge, which may play a role in charge-carrier recombination. While the observation of shallow or resonant transition levels for the V_{MA}, V_{SCN}, and V_I defects is consistent with previous reports in the literature, our prediction of an ultradeep V_{Pb} is at odds to the work of Xiao et al. who reported a shallower $-2/-1$ level 0.44 eV above the VBM [31]. The cause of this discrepancy may be due to the different computational parameters used in our calculations, namely, we have employed a larger supercell (450 vs. 400 atoms), tighter convergence criteria during geometry relaxations (0.01 vs. 0.05 eV $\mathring{A}^{-1}$) and our use of the PBEsol exchange–correlation functional versus PBE.

Fig. 5.5 Charge-state transition level diagram for a range of intrinsic vacancy defects in $(CH_3NH_3)_2Pb(SCN)_2I_2$, calculated using PBEsol. Red bands with filled circles denote donor defects, green bands with open circles denote acceptor defects

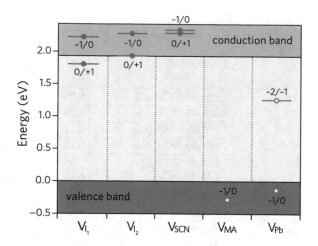

5.4 MAPSI Structured Analogues

The confirmation of MAPSI as a suitable absorber material for photovoltaic top cells opens up some fundamental questions. Similarly to the ABX_3 structured perovskites, which can be tuned on the A, B and X sites, can MAPSI also function as a parent compound to a range of analogues possessing the same structural motif? Chemically, it should be viable to replace Pb with other 2+ cations such as Sn, I with Cl and Br, and SCN with other pseudohalides such as OCN and SeCN. We have therefore considered a total of 17 analogue compounds, comprising all possible compositions of $(CH_3NH_3)_2MPs_2X_2$ (where M = Sn, Pb; Ps = OCN, SCN, SeCN; and X = Cl, Br, I). As all replacement elements are isoelectronic with their counterparts, the geometric and electronic structure of MAPSI should remain relatively unperturbed. Indeed, when structural relaxations were performed using the PBEsol functional, all compounds retained the orthorhombic $Pnm2_1$ space group of MAPSI, with only subtle distortions of the local bonding observed.

5.4.1 Stabilities

In order to check the stability of the analogues, we have computed the enthalpy of decomposition using PBEsol, for three routes comparable to those tested for MAPSI. For $(CH_3NH_3)_2MPs_2X_2$ (where M = Sn, Pb; Ps = OCN, SCN, SeCN; and X = Cl, Br, I) the decomposition paths investigated were:

Table 5.2 Enthalpy of decomposition, $\Delta_d H$, with respect to Eqs. (5.5), (5.6), and (5.7), for $(CH_3NH_3)_2 MPs_2 X_2$, where M = Sn, Pb; Ps = OCN, SCN, SeCN; and X = Cl, Br, I, calculated using PBEsol

	Compound	$\Delta_d H$ (eV)		
		Eq. (5.5)	Eq. (5.6)	Eq. (5.7)
Pb	$MA_2Pb(OCN)_2Cl_2$	0.05	2.66	0.72
	$MA_2Pb(SCN)_2Cl_2$	0.08	2.99	0.90
	$MA_2Pb(SeCN)_2Cl_2$	0.11	3.23	1.03
	$MA_2Pb(OCN)_2Br_2$	0.10	0.63	0.85
	$MA_2Pb(SeCN)_2Br_2$	0.25	1.29	1.25
	$MA_2Pb(OCN)_2I_2$	0.13	1.41	0.79
	$MA_2Pb(SCN)_2I_2$	0.38	1.97	1.20
	$MA_2Pb(SeCN)_2I_2$	0.41	2.21	1.33
Sn	$MA_2Sn(OCN)_2Cl_2$	4.09	1.62	2.78
	$MA_2Sn(SCN)_2Cl_2$	0.16	2.09	1.04
	$MA_2Sn(SeCN)_2Cl_2$	0.11	2.31	1.13
	$MA_2Sn(OCN)_2Br_2$	4.19	1.66	2.87
	$MA_2Sn(SCN)_2Br_2$	0.22	2.09	1.10
	$MA_2Sn(SeCN)_2Br_2$	0.18	2.31	2.31
	$MA_2Sn(OCN)_2I_2$	4.19	1.46	2.85
	$MA_2Sn(SCN)_2I_2$	0.34	2.01	1.20
	$MA_2Sn(SeCN)_2I_2$	0.32	2.25	1.31

$$(CH_3NH_3)_2 MPs_2 X_2 \rightarrow 2\,CH_3NH_3 X + MPs_2 \tag{5.5}$$

$$(CH_3NH_3)_2 MPs_2 X_2 \rightarrow 2\,CH_3NH_3 Ps + MX_2 \tag{5.6}$$

$$(CH_3NH_3)_2 MPs_2 X_2 \rightarrow CH_3NH_3 Ps + \tfrac{1}{2}CH_3NH_3 MX_3$$
$$+ \tfrac{1}{2}CH_3NH_3 X + \tfrac{1}{2}MPs_2 \tag{5.7}$$

The enthalpies of decomposition, $\Delta_d H$, for all compounds and decomposition routes are shown in Table 5.2. The enthalpies of decomposition were positive across the board, suggesting that all analogues should be stable against phase separation.

5.4.2 Electronic Properties

The fundamental direct and indirect band gaps for all analogues are provided in Table 5.3. We note that, as in MAPSI, the band gaps are likely to be underestimated relative to the room temperature optical band gaps. In contrast to MAPSI, where the fundamental band gap is direct, most of the analogues posses slightly indirect gaps. The difference between direct and indirect transitions is, however, small (between 10 and 40meV), and will therefore have a limited effect on the optical absorption.

Table 5.3 Electronic properties of $(CH_3NH_3)_2MPs_2X_2$, where M = Sn, Pb; Ps = OCN, SCN, SeCN; and X = Cl, Br, I, calculated using HSE06+SOC. Indirect and direct band gaps denoted by E_g^{ind} and E_g^{dir}, respectively. Hole and electron effective masses are given by m_h^* and m_e^*. IP and EA stand for ionisation potential and electron affinity, respectively. $\perp$ and $\parallel$ indicate properties perpendicular and parallel to the two-dimensional perovskite sheets, respectively

	Compound	E_g^{ind} (eV)	E_g^{dir} (eV)	$m_h^{*\perp}$ (m_0)	$m_h^{*\parallel}$ (m_0)	$m_e^{*\parallel}$ (m_0)	IP (eV)	EA (eV)
Pb	$MA_2Pb(OCN)_2Cl_2$	3.47	3.51	41.10	0.59	0.42	7.14	3.67
	$MA_2Pb(SCN)_2Cl_2$	3.03	3.05	40.00	3.07	0.50	6.50	3.94
	$MA_2Pb(SeCN)_2Cl_2$	3.04	3.05	38.89	14.09	0.59	5.68	3.93
	$MA_2Pb(OCN)_2Br_2$	2.56	2.58	–	0.34	0.28	6.38	3.34
	$MA_2Pb(SCN)_2Br_2$	2.30	2.32	36.17	0.44	0.25	6.03	3.73
	$MA_2Pb(SeCN)_2Br_2$	2.31	2.33	–	1.89	0.29	5.55	3.76
	$MA_2Pb(OCN)_2I_2$	1.75	1.76	–	0.22	0.18	6.27	3.23
	$MA_2Pb(SCN)_2I_2$	1.79	1.79	–	0.31	0.20	5.91	3.60
	$MA_2Pb(SeCN)_2I_2$	1.82	1.85	–	0.86	0.19	5.51	3.68
Sn	$MA_2Sn(OCN)_2Cl_2$	4.03	4.05	39.52	1.36	0.63	6.67	2.64
	$MA_2Sn(SCN)_2Cl_2$	3.74	3.74	38.74	20.42	0.96	5.94	2.88
	$MA_2Sn(SeCN)_2Cl_2$	3.56	3.57	37.59	38.66	0.86	5.13	3.22
	$MA_2Sn(OCN)_2Br_2$	3.06	3.08	–	0.36	0.43	6.06	2.32
	$MA_2Sn(SCN)_2Br_2$	2.98	2.98	–	2.15	0.63	5.67	2.69
	$MA_2Sn(SeCN)_2Br_2$	2.96	2.96	–	3.85	0.64	4.79	3.12
	$MA_2Sn(OCN)_2I_2$	1.90	1.92	–	0.19	0.32	5.90	2.33
	$MA_2Sn(SCN)_2I_2$	1.67	1.69	–	0.51	0.32	5.64	2.67
	$MA_2Sn(SeCN)_2I_2$	1.89	1.89	–	0.84	0.38	4.94	3.06

The size of the band gaps is controlled by the X site anion: when moving down group 17, the band gap is reduced from $\sim$3.0 eV (Cl) to $\sim$2.4 eV (Br) and $\sim$1.8 eV (I). An examination of the density of states of MAPSI can rationalise this behaviour: the upper valence band is composed almost entirely of halide p states, with the conduction band dominated by Pb p states. Accordingly, the valence band character causes the band gap reduction seen down the series, in line with the binding energy of the halide p orbitals. In comparison, the pseudohalide has a much smaller effect on the band gap, with the thiocyanate and selenocyanate showing only slightly smaller band gaps than their cyanate counterparts.

The trend in band gaps between metals is more pronounced, with the tin analogues possessing band gaps an average of 0.40 eV larger than the corresponding lead materials. This behaviour reflects the strength of spin–orbit coupling on the lead valence electrons, which acts to increase the binding energy of the Pb $5p$ orbitals and lower the conduction band minimum. The aforementioned trends of the pseudohalides and metals are not seen in the band gaps of the iodides, which show little correlation to composition. In general, the iodides all possess near-direct band gaps

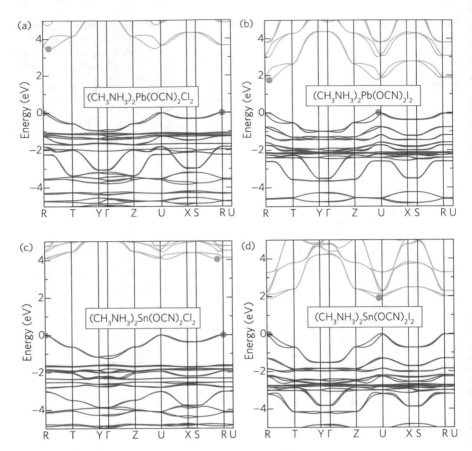

Fig. 5.6 HSE06+SOC calculated band structure of MAPSI-structured cyanate compounds: **a** $(CH_3NH_3)_2Pb(OCN)_2Cl_2$, **b** $(CH_3NH_3)_2Pb(OCN)_2I_2$, **c** $(CH_3NH_3)_2Sn(OCN)_2Cl_2$, and **d** $(CH_3NH_3)_2Sn(OCN)_2I_2$. The valence band maximum is set to 0 eV. Valence and conduction bands indicated by blue and orange lines, respectively.

in the ideal range for photovoltaic top cells (1.69–1.92 eV), even when considering the HSE06+SOC results are likely underestimated compared to the experimental gaps.

Spin–orbit coupling was found to have a dramatic effect on the band gaps of all materials (full results provided in Table A.1 of Appendix A). The reduction in the band gaps of the lead analogues (∼0.6–0.9 eV) was significantly larger than in their tin counterparts (∼0.1–0.2 eV), as spin–orbit coupling interactions scale approximately as Z^4, where Z is the atomic mass. These results stress the need to properly account for relativistic effects when modelling lead and tin hybrid perovskite-type materials.

To illustrate the effect of the halide on electronic properties, a selection of band structures for the tin and lead cyanate analogues are provided in Fig. 5.6 (the full set of band structures for each composition can be found in Appendix A). Moving down the halides, the reduction in band gap is apparent, with the width of the valence and conduction bands remaining roughly constant for all compositions. In both the lead and tin chlorides, the valence band maximum is found at the R point ($\frac{\bar{1}}{2}, \frac{1}{2}, \frac{1}{2}$) and shifts slightly to the side of either the R or U ($0, \frac{1}{2}, \frac{1}{2}$) points for the bromides, due to greater Rashba splitting in the upper valence bands. The conduction band minimum position does not appear to follow any trends in composition and is found close to the R or U points, depending on the asymmetry in Rashba splitting. The reduction in band gap occurs with a concomitant reduction in the band edge effective masses. Considering only the two directions parallel to the 2D perovskite sheets (along [100] and [001], indicated by $m^{*\parallel}$), the hole and electron effective masses are $m_h^{*\parallel} = 0.59\,m_0, 0.34\,m_0$, and $0.22\,m_0$ and $m_e^{*\parallel} = 0.42\,m_0, 0.28\,m_0, 0.18\,m_0$ for Cl, Br, and I lead cyanates, respectively. In contrast, the effective masses for the directions perpendicular to the inorganic perovskite layers (along [010]) are considerably larger ($\sim$40$\,m_0$), as expected due to the 2D structure, which results in limited interactions across the layers.

To illustrate the effect of the pseudohalide on electronic properties, a selection of band structures for the tin bromide and iodide analogues are provided in Fig. 5.7 (the full set of band structures for each composition can be found in Appendix A). Moving from OCN to SCN and SeCN, the valence bandwidth is significantly reduced, with the conduction bandwidth showing a similar, albeit lesser, trend. We note that this effect is stronger in the chlorides and bromides than for the iodides. The pseudohalide plays a role on the in-plane effective masses, with the SCN and SeCN analogues showing masses almost twice as light as those for the OCN compounds. Furthermore, the choice of pseudohalide appears to play a role in controlling the anisotropy in the Rashba splitting of the conduction band minimum. Indeed, while the spin-split pockets appear almost symmetrical in the thiocyanates and seleno-cyanates, the spin pockets of the cyanates show significant asymmetry. Considering the lower conduction band is dominated by metal p states, this behaviour is likely due to distortion in the local symmetry around the metal site. This is consistent with analysis of the crystal structures, which reveals greater off-centring of the metal cation within the MX_4Ps_2 octahedra for the cyanate materials, than in the thiocyanate and selenocyanate analogues. This is likely due to the formation of a stereochemically active lone-pair of electrons in the metal s^2 orbitals, which, based on the revised lone-pair model [32, 33], will be most prevalent for cyanate and reduced for the thiocyanate and selenocyanate analogues.

Analysis of these results reveals a trend; while the chlorides and bromides show reduced band gaps and narrower valence bandwidths when moving down the pseudohalides, the iodides do not show the same pattern. Investigation of the relative energies of the pseudohalide and halide p orbitals reveals the origin of this discrepancy. Here we are chiefly concerned with the energies of the chalcogenide p orbitals (rather than the energies of the N and C states), due to their direct bonding to the metal site. Considering the OCN analogues: the upper valence band is dominated by halide p and metal s states, with the O p orbitals present $\sim$1–2 eV deeper into the valence band (Fig. 5.8a). In these materials, the band gaps, valence bandwidths and

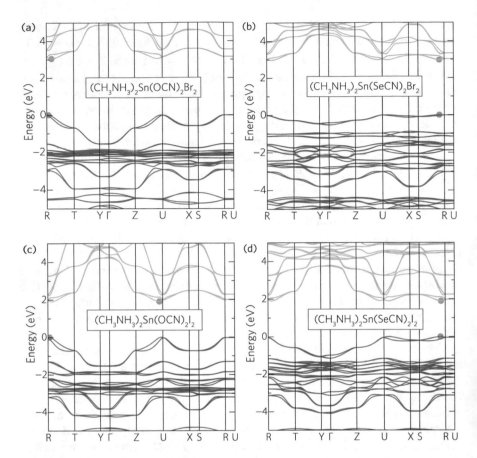

Fig. 5.7 HSE06+SOC calculated band structure of MAPSI-structured bromide compounds: **a** $(CH_3NH_3)_2Sn(OCN)_2Br_2$, **b** $(CH_3NH_3)_2Sn(SeCN)_2Br_2$, **c** $(CH_3NH_3)_2Sn(OCN)_2Br_2$, and **d** $(CH_3NH_3)_2Sn(SeCN)_2Br_2$. The valence band maximum is set to 0 eV. Valence and conduction bands indicated by blue and orange lines, respectively

hole effective masses are solely determined by the halide, resulting in smaller effective masses in the diffuse iodides than the more localised bromides and chlorides. As the S $3p$ and Se $4p$ states are higher in energy (10.4 eV and 9.8 eV, respectively) [34] than the chloride and bromide p states (13.0 and 11.8 eV) [34], the valence band maximum in the thiocyanates and selenocynates shows considerably greater chalcogen contribution. As such, the dispersion at the valence band edge is reduced, due to the greater localisation of these states (Fig. 5.8b). In the iodides, however, the reduced binding energy of the I $5p$ orbitals (10.4 eV) [35]) enables them to hybridise with the S and Se states, resulting in a consistent band gap regardless in the choice of pseudohalide, with limited effects seen on the valence bandwidth.

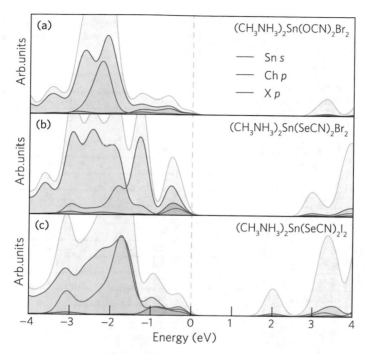

Fig. 5.8 Projected density of states for **a** $(CH_3NH_3)_2Sn(OCN)_2Br_2$, **b** $(CH_3NH_3)_2Sn(SeCN)_2Br_2$, and **c** $(CH_3NH_3)_2Sn(SeCN)_2I_2$, highlighting the relative contributions of the chalcogenide (Ch) and halide (X) to the upper valence band. The valence band maximum is set to 0 eV and is indicated by a dashed line. The contributions of the N, C, and Sn p orbitals have been omitted for clarity

5.4.3 Dielectric Constants

The PBEsol calculated dielectric constants are provided in Table 5.4. For most compounds, the dielectric constants parallel to the perovskite sheets are larger and show greater variation ($\varepsilon_r^{\parallel} = 6.1$–$21.9$) than for the directions perpendicular to the sheets ($\varepsilon_r^{\perp} = 6.4$–$12.5$). The tin compounds generally show smaller dielectric constants than their lead analogues, due to the reduced polarisability of the Sn^{2+} cation. Considering just the lead-based materials, the halide plays a clear role on the magnitude of $\varepsilon_r^{\parallel}$, with the iodides and bromides showing larger dielectric constants ($\sim$20 and $\sim$18, respectively) than the chlorides ($\sim$10). These trends are broadly reproduced in the tin analogues, with the iodides again showing the largest in-plane dielectric constants ($\sim$12–20), in comparison to much smaller constants for the bromides ($\sim$8) and chlorides ($\sim$7). The relatively large dielectric constants of the iodides are particularly promising for photovoltaic applications, where increased screening of charge will limit the effects of scattering and recombination by charged defects and reduce defect binding energies.

Table 5.4 Dielectric constants (ε_r) of $(CH_3NH_3)_2MPs_2X_2$, where M = Sn, Pb; Ps = OCN, SCN, SeCN; and X = Cl, Br, I, calculated using HSE06+SOC. $\perp$ and $\parallel$ indicate properties perpendicular and parallel to the 2D perovskite sheets, respectively

	Compound	$\varepsilon_r^{\perp}$	$\varepsilon_r^{\parallel}$
Pb	$MA_2Pb(OCN)_2Cl_2$	7.2	11.4
	$MA_2Pb(SCN)_2Cl_2$	6.6	19.5
	$MA_2Pb(SeCN)_2Cl_2$	6.4	19.0
	$MA_2Pb(OCN)_2Br_2$	8.4	9.1
	$MA_2Pb(SCN)_2Br_2$	8.3	19.6
	$MA_2Pb(SeCN)_2Br_2$	7.4	16.8
	$MA_2Pb(OCN)_2I_2$	9.3	8.5
	$MA_2Pb(SCN)_2I_2$	9.0	21.9
	$MA_2Pb(SeCN)_2I_2$	9.2	17.7
Sn	$MA_2Sn(OCN)_2Cl_2$	9.3	6.1
	$MA_2Sn(SCN)_2Cl_2$	6.9	7.0
	$MA_2Sn(SeCN)_2Cl_2$	7.0	11.7
	$MA_2Sn(OCN)_2Br_2$	12.5	7.2
	$MA_2Sn(SCN)_2Br_2$	8.3	8.4
	$MA_2Sn(SeCN)_2Br_2$	8.5	20.2
	$MA_2Sn(OCN)_2I_2$	9.3	6.9
	$MA_2Sn(SCN)_2I_2$	8.7	7.8
	$MA_2Sn(SeCN)_2I_2$	8.8	13.9

5.4.4   Band Alignments

When designing a photovoltaic device, it is essential to achieve close band alignment of the absorber material with the transparent conducting oxide and hole transporting material, to maximise the achievable open-circuit voltage. The band alignments are also crucial in understanding the dopability of a material—a measure of the ease of n or p type doping. To aid the experimental realisation of high efficiency photovoltaics based on MAPSI and its analogues, we have calculated the band alignments of $(CH_3NH_3)_2MPs_2X_2$, (where M = Sn, Pb; Ps = OCN, SCN, SeCN; and X = Cl, Br, I) using HSE06 with a correction for the HSE06+SOC band gap (Fig. 5.9).

It is immediately clear that the electron affinity (EA)—the distance from the conduction band minimum to the vacuum level—shows significantly less variability than the ionisation potential (IP)—the distance from the valence band maximum to the vacuum level. Accordingly, the magnitude of the band gap is controlled through the depth of the valence band maximum, and is therefore determined mainly by the choice in halide. In agreement with the previous density-of-states analysis, the electron affinity and ionisation potential remain relatively unchanged across comparable (i.e. compounds possessing the same halide and metal composition) SCN and SeCN analogues. Overall, the electron affinities and ionisation potentials of the tin com-

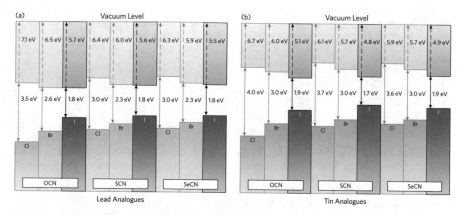

Fig. 5.9 Band alignment of **a** $(CH_3NH_3)_2PbPs_2X_2$ and **b** $(CH_3NH_3)_2SnPs_2X_2$ (where Ps = OCN, SCN, SeCN; and X = Cl, Br, I), relative the vacuum level. Electron affinities and band gaps indicated by dashed and solid lines, respectively

pounds are slightly smaller than their lead counterparts, due to the greater relativistic stabilisation effects seen on the Pb s and p orbitals. The greater electron affinities of the lead analogues is expected to allow for increased ease of n-type doping, whereas the larger ionisation potential indicates enhanced stability toward oxidation.

The calculated IP of MAPSI (5.6 eV) is in good agreement with the experimentally determined value (5.7 eV) [36]. This is close to the ionisation potential of MAPI (5.5–5.7 eV) [37, 38], as expected due to the similar valence band composition dominated by Pb s and I p states. As such, devices based on the lead iodide analogues are expected to form efficient band alignments with spiro-OMeTAD (IP = 5.2 eV) [38], a common hole transport layer used in the best performing hybrid halide devices. Additionally, the electron affinity of the lead iodide analogues (3.7–3.9 eV), suggests the absorbers will form favourable alignments with fluorine-doped tin oxide (FTO, EA = 4.4 eV) [39], a cheap and commercially available transparent conducting oxide.

5.4.5 Defect Chemistry

The defect chemistry of all analogues is broadly similar to that of MAPSI (the transition level diagrams for all compounds are provided in Fig. A.1 of Appendix A). The transition level diagram of $MA_2Pb(OCN)_2I_2$ is shown in Fig. 5.10a. Of the n-type defects, the iodide vacancies possess relatively shallow $0/+1$ transition levels (0.16 and 0.08 eV below the conduction band minimum), with the V_{OCN} defect level resonant in the conduction band. For the p-type defects, V_{MA} is again found to be resonant in the valence band, due to the weak interactions between the methylammonium and perovskite lattice. Similar to MAPSI, the V_{Pb} $-2/-1$ defect level is ultra-deep in the band gap and therefore may contribute to charge-carrier recombination. Moving

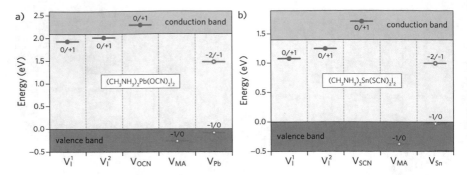

Fig. 5.10 Calculated charge-state transition level diagrams for a range of intrinsic vacancy defects in **a** $(CH_3NH_3)_2Pb(OCN)_2I_2$, and **b** $(CH_3NH_3)_2Sn(SCN)_2I_2$ calculated at the PBEsol level. Red bands with filled circles indicate donor defects, green bands with open circles indicate acceptor defects

from lead to tin, the defect chemistry remains largely unchanged, as indicated by the transition level diagram of $MA_2Sn(SCN)_2I_2$ (Fig. 5.10b). The main difference is found in the iodide vacancies, which possess defect states that are deeper into the gap (0.15 and 0.32 eV below the conduction band minimum). This can be rationalised by the band alignments, which show the tin analogues possess smaller electron affinities than their lead counterparts. As such, it is likely the iodide vacancy transition levels remain fixed in energy, with the shift in energy of the conduction band minimum causing the states to lie further from the conduction band edge.

5.5   Conclusions

In this chapter, we have investigated $(CH_3NH_3)_2Pb(SCN)_2I_2$ (MAPSI) and its analogues as potential solar absorber materials. We find that MAPSI possesses an electronic structure suitable for a photovoltaic top cell, with a band gap of $\sim$1.79 eV (albeit slightly underestimated compared to the experimental gap of $\sim$2.1 eV), small charge-carrier effective masses within the perovskite-type layers, and enhanced stability against decomposition compared with the prototypical hybrid perovskite, $CH_3NH_3PbI_3$ (MAPI). Our results suggest that MAPSI maintains a smaller band gap, when compared to other two-dimensional absorbers due to the presence of SCN states at the top of the valence band maximum. Furthermore, its intrinsic defect chemistry shows few deep defect states that could contribute to carrier recombination.

We have further investigated the potential for MAPSI to act as a parent compound to a range of analogues with the same crystal structure, thereby allowing for a material with three degrees of property tuning. All materials in the $(CH_3NH_3)_2MPs_2X_2$ series (where M = Pb, Sn; Ps = OCN, SCN, SeCN; and X = Cl, Br, I) were found to maintain the MAPSI structural motif, with all showing stability against phase separation. We demonstrated that the electronic properties of the series are dictated primarily by the choice in halide, with several compounds possessing band gaps suitable for photovoltaic applications. Band alignment calculations reveal MAPSI shows a similar ionisation potential and electron affinity to MAPI, indicating that hole and electron contact materials commonly used in the production of hybrid halide devices, will also perform well for MAPSI and its analogues. Overall, the iodide-based analogues show the most promise for solar devices, due to their ideal band gaps and relatively large dielectric constants. Due to concerns over the toxicity of lead, $(CH_3NH_3)_2Sn(SCN)_2I_2$ is a particularly exciting addition to the perovskite family, based on its potential as a stable and efficient lead-free photovoltaic.

Notes

The work in this chapter formed the basis for two publications:

1. A. M. Ganose, C. N. Savory and D. O. Scanlon, $(CH_3NH_3)_2Pb(SCN)_2I_2$: a more stable structural motif for hybrid halide photovoltaics?, *Journal of Physical Chemistry Letters*, **6**, 4594 (2015).
2. A. M. Ganose, C. N. Savory and D. O. Scanlon, $(CH_3NH_3)_2PbI_2(SCN)_2$ analogues for photovoltaic applications, *Journal of Materials Chemistry A*, **5**, 7845–7853 (2017).

The optimised crystal structures, for all compounds discussed in this chapter, are provided in an online repository [40].

References

1. Ganose AM, Savory CN, Scanlon DO (2017) Electronic and defect properties of $(CH_3NH_3)_2Pb(SCN)_2I_2$ analogues for photovoltaic applications. J Mater Chem A 5:7845–7853
2. Ganose AM, Savory CN, Scanlon DO (2015) $(CH_3NH_3)_2Pb(SCN)_2I_2$: a more stable structural motif for hybrid halide photovoltaics? J Phys Chem Lett 6:4594–4598
3. Chen C, Li W, Zhou Y, Chen C, Luo M, Liu X, Zeng K, Yang B, Zhang C, Han J, Tang J (2015) Optical properties of amorphous and polycrystalline Sb_2Se_3 thin films prepared by thermal evaporation. Appl Phys Lett 107:043905
4. Halder A, Chulliyil R, Subbiah AS, Khan T, Chattoraj S, Chowdhury A, Sarkar SK (2015) Pseudohalide (SCN^-)-Doped MAPbI3 perovskites: a few surprises. J Phys Chem Lett 6:3483–3489
5. Jiang Q, Rebollar D, Gong J, Piacentino EL, Zheng C, Xu T (2015) Pseudohalide-induced moisture tolerance in perovskite $CH_3NH_3Pb(SCN)_2I$ thin films. Angew Chem Int Ed 54:7617–7620
6. Daub M, Hillebrecht H (2015) Synthesis, single-crystal structure and characterization of $(CH_3NH_3)_2Pb(SCN)_2I_2$. Angew Chem 127:11168–11169
7. Iwadate Y, Kawamura K, Igarashi K, Mochinaga J (1982) Effective ionic radii of nitrite and thiocyanate estimated in terms of the Boettcher equation and the Lorentz-Lorenz equation. J Phys Chem 86:5205–5208

8. Kim H-S, Park N-G (2014) Parameters affecting I-V hysteresis of $CH_3NH_3PbI_3$ Perovskite solar cells: effects of perovskite crystal size and mesoporous TiO_2 layer. J Phys Chem Lett 5:2927–2934
9. Snaith HJ, Abate A, Ball JM, Eperon GE, Leijtens T, Noel NK, Stranks SD, Wang JT-W, Wojciechowski K, Zhang W (2014) Anomalous hysteresis in Perovskite solar cells. J Phys Chem Lett 5:1511–1515
10. Niu G, Li W, Meng F, Wang L, Dong H, Qiu Y (2014) Study on the stability of $CH_3NH_3PbI_3$ films and the effect of post-modification by aluminum oxide in all-solid-state hybrid solar cells. J Mater Chem A 2:705–710
11. Chen Y, Li B, Huang W, Gao D, Liang Z (2015) Efficient and reproducible $CH_3NH_3PbI_{3-x}(SCN)_x$ perovskite based planar solar cells. Chem Commun 51:11997–11999
12. Xiao Z, Meng W, Saparov B, Duan H-S, Wang C, Feng C, Liao W-Q, Ke W, Zhao D, Wang J, Mitzi DB, Yan Y (2016) Photovoltaic properties of two-dimensional $(CH_3NH_3)_2Pb(SCN)_2I_2$ Perovskite: a combined experimental and density-functional theory study. J Phys Chem Lett 7:1213–1218
13. Umeyama D, Lin Y, Karunadasa HI (2016) Red-to-black piezochromism in a compressible Pb-I-SCN layered perovskite. Chem Mater 28:3241–3244
14. Younts R, Duan H-S, Gautam B, Saparov B, Liu J, Mongin C, Castellano FN, Mitzi DB, Gundogdu K (2017) Efficient generation of long-lived triplet excitons in 2D hybrid perovskite. Adv Mater 1–7
15. Tanaka K, Kondo T (2003) Bandgap and exciton binding energies in lead-iodide-based natural quantum-well crystals. Sci Technol Adv Mater 4:599–604
16. Pulay P (1969) Ab initio calculation of force constants and equilibrium geometries in polyatomic molecules: I. Theory Mol Phys 17:197–204
17. Wei S-H, Zunger A (1998) Calculated natural band offsets of all II-VI and III-V semiconductors: chemical trends and the role of cation D orbitals. Appl Phys Lett 72:2011–2013
18. Wei S-H, Zunger A (1988) Role of metal d states in II-VI semiconductors. Phys Rev B 37:8958
19. Williams DJ, Daemen L, Vogel S, Proffen T (2007) Temperature dependence of the crystal structure of α-AgSCN by powder neutron diffraction. J Appl Crystallogr 40:1039–1043
20. Hendon CH, Yang RX, Burton LA, Walsh A (2015) Assessment of polyanion (BF_4^- And PF_6^-) substitutions in hybrid halide perovskites. J Mater Chem A 3:9067–9070
21. Azarhoosh P, Frost JM, McKechnie S, Walsh A, van Schilfgaarde M (2016) Relativistic origin of slow electron-hole recombination in hybrid halide perovskite solar cells. APL Mater 4:091501
22. Frost JM, Butler KT, Walsh A (2014) Molecular ferroelectric contributions to anomalous hysteresis in hybrid perovskite solar cells. APL Mater 2:081506
23. Fernandes P, Salomé P, Da Cunha A (2009) Precursors' order effect on the properties of sulfurized Cu_2ZnSnS_4 thin films. Semicond Sci Technol 24:105013
24. Mosconi E, Amat A, Nazeeruddin MK, Grätzel M, De Angelis F (2013) First-principles modeling of mixed halide organometal perovskites for photovoltaic applications. J Mater Chem C 117:13902–13913
25. Butler KT, Frost JM, Walsh A (2015) Ferroelectric materials for solar energy conversion: photoferroics revisited. Energy Environ Sci 8:838–848
26. Guarnera S, Abate A, Zhang W, Foster JM, Richardson G, Petrozza A, Snaith HJ (2015) Improving the long-term stability of Perovskite solar cells with a porous Al_2O_3 buffer layer. J Phys Chem Lett 6:432–437
27. Niu G, Guo X, Wang L (2015) Review of recent progress in chemical stability of Perovskite solar cells. J Mater Chem A 3:8970–8980
28. Burschka J, Pellet N, Moon S-J, Humphry-Baker R, Gao P, Nazeeruddin MK, Grätzel M (2013) Sequential deposition as a route to high-performance Perovskite-sensitized solar cells. Nature 499:316–319
29. Pisoni A, Jaćimović J, Barišić OS, Spina M, Gaál R, Forró L, Horváth E (2014) Ultra-low thermal conductivity in organic-inorganic hybrid perovskite $CH_3NH_3PbI_3$. J Phys Chem Lett 5:2488–2492

30. Zhang Y-Y, Chen S, Xu P, Xiang H, Gong X-G, Walsh A, Wei S-H (2018) Intrinsic instability of the hybrid halide perovskite semiconductor $CH_3NH_3PbI_3$. Chin Phys Lett 35:036104
31. Xiao Z, Meng W, Wang J, Yan Y (2016) Defect properties of the two-dimensional $(CH_3NH_3)_2Pb(SCN)_2I_2$ perovskite: a density-functional theory study. Phys Chem Chem Phys 18:25786–25790
32. Payne DJ, Egdell RG, Walsh A, Watson GW, Guo J, Glans PA, Learmonth T, Smith KE (2006) Electronic origins of structural distortions in post-transition metal oxides: experimental and theoretical evidence for a revision of the lone pair model. Phys Rev Lett 96:157403
33. Walsh A, Payne DJ, Egdell RG, Watson GW (2011) Stereochemistry of post-transition metal oxides: revision of the classical lone pair model. Chem Soc Rev 40:4455–4463
34. Others et al (1989) CRC handbook of chemistry and physics, vol 1990. CRC Press, Boca Raton FL
35. James AM, Lord MP (1992) Macmillan's chemical and physical data. Macmillan London
36. Liu J, Shi J, Li D, Zhang F, Li X, Xiao Y, Wang S (2016) Molecular design and photovoltaic performance of a novel thiocyanate-based layered organometal perovskite material. Synth Met 215:56–63
37. Frost JM, Butler KT, Brivio F, Hendon CH, Van Schilfgaarde M, Walsh A (2014) Atomistic origins of high-performance in hybrid halide Perovskite solar cells. Nano Lett 14:2584–2590
38. Green MA, Ho-Baillie A, Snaith HJ (2014) The emergence of Perovskite solar cells. Nat Photon 8:506–514
39. Helander M, Greiner M, Wang Z, Tang W, Lu Z (2011) Work function of fluorine doped tin oxide. J Vac Sci Technol A 29:011019
40. https://github.com/SMTG-UCL/MAPSI . Accessed 27 Sept 2014

Chapter 6
Vacancy-Ordered Double Perovskites

6.1 Introduction

The[1] tin-based hybrid perovskite, $MASnI_3$, and its inorganic counterpart, $CsSnI_3$, have received significant attention as lead-free alternatives for perovskite photovoltaics [2–6]. While the efficiencies of devices containing these compounds has reached over 9.0% in recent years, [7] their extremely poor long-term stability— worse, even, than their lead analogues—presents a serious challenge that remains unresolved [8–10]. This instability stems from the facile oxidation of Sn^{2+} to Sn^{4+}, which occurs readily in heat, light or contact with moisture [11, 12]. Accordingly, significant work will be required if these materials are to find viable practical applications [13].

The broad diversity in composition and structure exhibited by the perovskite family (formula ABX_3), provides a number of routes to circumvent these issues. For example, it is possible to produce double halide perovskites (with the formula $A_2B'B''X_6$), which show rock-salt ordering of the B' and B'' metal cations due to size and charge differences [14–16]. Unfortunately, recent work has shown most lead-free double perovskites will exhibit indirect band gaps, owing to a mismatch in angular momentum of the frontier electronic orbitals [17]. While this issue can be avoided through employing indium or thallium, the resulting compounds are either thermodynamically unstable or extremely toxic [17].

The related vacancy-ordered double perovskites can be produced by replacing one B-site cation with a vacancy, resulting in a material with the A_2BX_6 formula (or $A_2B\square X_6$, where $\square$ is a vacancy). They crystallise in the K_2PtCl_6 structure (termed antifluorite), containing isolated $[BX_6]^{2-}$ octahedra. The structure is charge balanced by A^+ cations residing in the cuboctahedral void, formed by the 12 nearest-neighbour X site anions (Fig. 6.1). Despite the loss of every other B-site cation, the A_2BX_6

---

[1]Parts of this chapter have been reproduced with permission from Ref. [1]—Published by the American Chemical Society.

© The Editor(s) (if applicable) and The Author(s), under exclusive license to Springer Nature Switzerland AG 2020
A. Ganose, *Atomic-Scale Insights into Emergent Photovoltaic Absorbers*, Springer Theses, https://doi.org/10.1007/978-3-030-55708-9_6

Cs
Sn
I

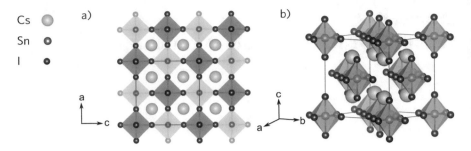

Fig. 6.1 Crystal structure of Cs_2SnI_6 viewed along **a** the [010] direction, **b** a three-dimensional view, highlighting the isolated $[SnI_6]^{2-}$ octahedral units. Sn, I, and Cs atoms are denoted by grey, purple, and turquoise spheres, respectively

perovskites demonstrate many similar properties to their ABX_3 counterparts, in part due to the retention of the close-packed anionic sublattice.

The inorganic Cs_2SnI_6 defect-perovskite has recently attracted attention as a stable alternative to $CsSnI_3$. Crystallising in the cubic $Fm\bar{3}m$ space group (Fig. 6.1), it possesses many properties ideal for photovoltaics; namely a band gap of ~1.3 eV, strong optical absorption over $10^5 cm^{-1}$ at 1.7 eV above the valence band edge, and intrinsic n-type electrical conductivity [18, 19]. Furthermore, Cs_2SnI_6 shows remarkably improved chemical stability in the presence of moisture and heat (stable up to 270 °C), due the formally 4+ oxidation state of Sn in the structure [20]. Remarkably, polycrystalline pellets of Cs_2SnI_6 have reportedly achieved carrier mobilities comparable to that of $CsSnI_3$ (μ = 310 and 585 $cm^2V^{-1}s^{-1}$, respectively) [8, 18]. A recent report on Cs_2SnI_6 thin films observed more modest electron mobilities of 3 $cm^2V^{-1}s^{-1}$, however, these results are still surprising considering the presence of regular B-site vacancies [19]. While a thorough understanding of the origins of conductivity in Cs_2SnI_6 remains limited, the formation of donor iodine vacancy defects, in addition to small electron effective masses, has been proposed [18, 19, 21].

The first devices containing the vacancy-ordered perovskites employed Cs_2SnI_6 as the hole transporting layer for dye-sensitised solar cells, achieving power conversion efficiencies of ~8% [18, 22]. Cs_2SnI_6 was subsequently incorporated as the absorber layer on TiO_2 and ZnO nanorods, achieving efficiencies of 0.96% and 0.86%, respectively [23, 24]. Work by Lee et al. on optimising the crystallinity of solution-processed Cs_2SnI_6 thin films, further enabled devices with efficiencies up to 1.47% [25]. The authors additionally explored members of the solid solution, $Cs_2SnI_xBr_{6-x}$, allowing for cells with efficiencies over 2.0% (for the composition $Cs_2SnI_4Br_2$), due to enhanced open-circuit voltage. Recent developments in producing Cs_2SnI_6 nanocrystals with tuneable optoelectronic properties, including effective masses a factor of 4 smaller than the bulk material, highlight the potential for further advancements in device engineering [26].

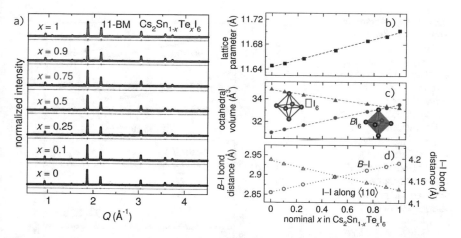

Fig. 6.2 Structural data for the solid-solution series, $Cs_2Sn_{1-x}Te_xI_6$. **a** High-resolution synchrotron X-ray diffraction data and Rietveld refinements, collected at room temperature. Black circles indicate collected data, coloured lines indicate the fit, and blue lines are the difference curves. **b–d** Structure parameters across the series, determined from high-resolution synchrotron X-ray diffraction. **b** The refined lattice parameters. Vegard's law interpolated from the end members is shown by a dashed blue line. **c** Average polyhedral volumes of the $\square I_6$ voids and $B\,I_6$ octahedra. As the volume of the $B\,I_6$ octahedra increase, the volume of the void regions decrease. **d** Average I-I and B-I bond distances, along the [110] direction. Dashed lines in **d** and **c** represent linear regressions. Data collected and plotted by Maughan et al. [1]

6.1.1 $Cs_2Sn_{1-x}Te_xI_6$ Solid Solutions

A proposed route to enhancing the carrier mobility of Cs_2SnI_6 is through substitution of Sn^{4+} (electron configuration: $[Kr]4d^{10}5s^0$) with Te^{4+} (electron configuration: $[Kr]4d^{10}5s^2$). Cs_2TeI_6 is isostructural with Cs_2SnI_6, however, the larger Te^{4+} is accommodated into the lattice through expansion of the $[BI]_6$ octahedra, reducing the volume of the inter-octahedral void. As the valence and conduction band edges are dominated by iodine character, the increased inter-octahedral I-I orbital overlap is expected to result in greater band dispersion and smaller effective masses. High-resolution synchrotron powder X-ray diffraction (SXRD) for the $Cs_2Sn_{1-x}Te_xI_6$ series, performed by our collaborators, Maughan et al., at Colorado State University, indicates tellurium substitution results in solid solution behaviour, with no distortion in the local coordination environment of each octahedron (Fig. 6.2a) [1]. The series shows a linear increase of lattice parameter upon tellurium substitution, following Vegards law, with a simultaneous decrease in interoctahedral I-I bond lengths (Fig. 6.2d).

As previously discussed, Cs_2SnI_6 is an intrinsic n-type semiconductor. This was confirmed by measurements performed by our collaborators, who found a resistivity of 13ωcm for cold-pressed polycrystalline pellets at T = 300 K, with Hall measurements indicating carrier concentrations of 5^16cm^{-1} and an electron mobilities of

Fig. 6.3 Optoelectronic
properties for the series
$Cs_2Sn_{1-x}Te_xI_6$. **a**
Dependence of the electrical
resistivity on temperature.
b Room temperature
resistivity. **c** Carrier mobility
(μ_e, open pink diamonds)
and carrier concentration
(n_e, filled blue triangles).
d The optical band gap,
derived from UV-visible
diffuse reflectance data. Data
collected and plotted by
Maughan et al. [1]

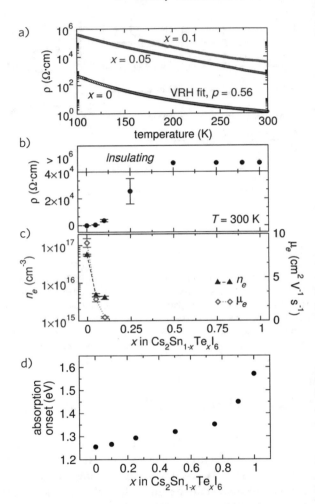

$8.6\,cm^2V^{-1}s^{-1}$ [1]. Unexpectedly, partial substitution of tin by tellurium, to form
the solid solution $Cs_2Sn_{1-x}Te_xI_6$, resulted in a significant increase in electrical resistivity (Fig. 6.3a). Substitution of 5% tellurium resulted in a 50-fold increase in resistivity, with samples containing more than 25% tellurium found to be too insulating
for accurate measurements. Hall measurements of the 5 and 10% tellurium members
of the solid solutions further revealed dramatically decreased carrier concentrations
and electron mobilities (Fig. 6.3b). Maughan et al. also observed a nonlinear change
in the measured optical band gap upon tellurium substitution (Fig. 6.3d), despite the
absence of any distortions in the local geometry, as determined by X-ray pair distribution function analysis [1]. As such, it is clear that a more detailed understanding of
the electronic, optical, and defect properties is essential to shed light on the behaviour
of the series.

6.2 Methodology

Calculations were performed using the Vienna *Ab initio* Simulation Package. A **k**-point mesh of Γ-centred $3 \times 3 \times 3$ and plane wave cutoff of 350 eV was found to converge the 9 atom primitive cells of Cs_2SnI_6 and Cs_2TeI_6 to within 1 meV/atom. During geometry optimisations, the cutoff was increased to 455 eV to avoid errors resulting from Pulay stress [27]. The structures were deemed converged when the forces totalled less than 10 meV Å^{-1}.

The HSE06 functional was employed for geometry relaxations. Electronic properties were calculated using HSE06 with the addition of spin–orbit coupling effects (HSE06+SOC). The Brillouin zone for the $Fm\bar{3}m$ space group, indicating the high-symmetry points explored in the band structure, is provided in Fig. 6.4 project crystal orbital Hamilton population (pCOHP) analysis was performed using the LOBSTER program, based on wavefunctions calculated using the HSE06 functional [28, 29].

For band alignment calculations, the core-level alignment approach of Wei and Zunger was employed [30], using a slab model with 35 Å of vacuum and a 45 Å thick slab. The slab was cleaved along the non-polar (101) surface, due to the absence of any dangling bonds. Due to the size of the model, which precluded the use of HSE06+SOC, band alignment calculations were performed using HSE06, with an explicit correction to the band gap and valence band maximum position taken from the HSE06+SOC calculated bulk.

Defect calculations were performed in a $3 \times 3 \times 3$ supercell containing 243 atoms, using the HSE06 functional and a single **k**-point at Γ. The defect energies were corrected to account for use of a finite-sized supercell, using the potential level alignment, band-filling and image-charge corrections described in Chap. 3.

Fig. 6.4 Brillouin zone of the $Fm\bar{3}m$ space group. Coordinates of the high symmetry points used for the band structures and effective masses: $\Gamma = (0, 0, 0)$; $X = (1/2, 0, 1/2)$; $L = (1/2, 1/2, 1/2)$; $W = (1/2, 1/4, 3/4)$

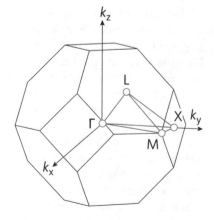

Table 6.1 HSE06 calculated lattice parameters and bond lengths for the conventional cells of Cs_2SnI_6 and Cs_2TeI_6. The lattice parameters and cation–anion interatomic distances are quoted in Å. The equilibrium crystal structures can be found online in a public repository [32]

Compound	a	d_{B-I}	d_{B-Cs}	d_{I-I}
Cs_2SnI_6	11.96	2.86	5.18	4.40
Cs_2TeI_6	11.98	2.94	5.19	4.32

6.3 Results

6.3.1 Geometric Structure

The lattice constants and structural parameters of Cs_2SnI_6 and Cs_2TeI_6, calculated using HSE06, are provided in Table 6.1. The results are in keeping with previous studies performed using HSE06 [21, 31]. The lattice parameters are slightly overestimated compared to experiment, by $\sim$2% in both cases. Moving from Cs_2SnI_6 to Cs_2TeI_6, the calculated lattice parameter increases by 0.02 Å; this is in good agreement to experiment, which shows a 0.05 Å increase in the a parameter. Furthermore, the change in the calculated B-I and I-I bond distances (0.08 Å in both cases), agrees well with the change seen in experiment (0.07 in both cases).

6.3.2 Electronic Properties

Some debate exists as to the nature and magnitude of the fundamental band gap in Cs_2SnI_6. Two previous experimental studies have reported a gap of 1.3 eV [19, 33], but differ on whether it is direct or indirect; whereas work on Cs_2SnI_6 thin films, suggested a direct gap of 1.6 eV [18]. The analysis performed by our co-workers, Maughan et al., suggests an optical gap of $\sim$1.25 eV for Cs_2SnI_6 [1]. Comparable analysis for Cs_2TeI_6, indicates an optical gap of $\sim$1.59 eV, [1] in agreement with the previously reported indirect gap of 1.5 eV [34].

The band structures of Cs_2SnI_6 and Cs_2TeI_6, calculated using HSE06 +SOC, are provided in Fig. 6.5. The band structure of Cs_2SnI_6 confirms it is a direct gap semiconductor, with a fundamental band gap of 0.97 eV occurring at the Γ point. This result is in reasonable agreement with GW0 calculations, which indicate a direct gap of 0.88 eV [22]. We note that the calculated band gap is noticeably smaller than the optical band gap seen in experiment ($\sim$1.3 eV)—a point to which we will return later. The band structure of Cs_2TeI_6 replicates experimental observations of an indirect band gap, however the calculated gap ($E_g^{ind} = 1.83$ eV) is larger than the experimental optical gap (1.59 eV). As observed in other tellurium-based materials, this difference may result from many-body effects that require higher order methods, such as configuration interactions, to correctly describe [35–38]. The conduction

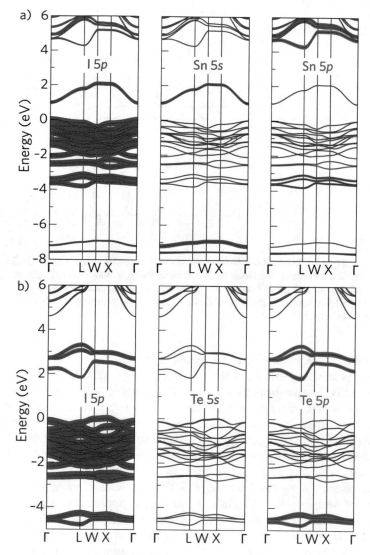

Fig. 6.5 Orbital projected band structures of **a** Cs_2SnI_6 and **b** Cs_2TeI_6, calculated using HSE06+SOC. The contributions from the I $5p$, Sn/Te $5s$, and Sn/Te $5p$ are shown by orange, blue, and green, respectively. The valence band maximum is set to 0 eV in all cases

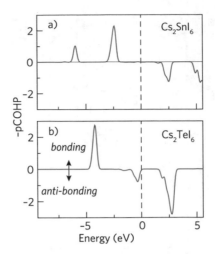

Fig. 6.6 Projected crystal orbital Hamilton population (pCOHP) analysis of **a** Cs$_2$SnI$_6$ and **b** Cs$_2$TeI$_6$, in which the density of states is partitioned with the sign denoting bonding or anti-bonding character, and the magnitude indicating the strength of the interaction. For each compound the pCOHP is averaged across all pairwise B-I interactions in the primitive cell. The valence band maximum is set to 0 eV

band minimum and valence band maximum of Cs$_2$TeI$_6$ occur at the X and L high-symmetry points, respectively, with the direct band gap (E$_g^{dir}$ = 2.05 eV) appearing at L. The effect of spin–orbit coupling on the electronic properties of Cs$_2$SnI$_6$ and Cs$_2$TeI$_6$ was tested, with the full results provided in Appendix B. In both cases, the relativistic renormalisation of the band gap occurs through raising of the valence band maximum by ~0.2 eV, with the conduction band remaining largely unaffected. The magnitude of this renormalisation indicates that inclusion of spin orbit effects is essential to accurately model these systems.

The band structures presented in Fig. 6.5, further include information on the orbital character of the bands. Both Cs$_2$SnI$_6$ and possess an upper valence band composed of I 5p states. In Cs$_2$SnI$_6$, the frontier valence orbitals are nonbonding, whereas for Cs$_2$TeI$_6$ there is a mixture of antibonding (Te 5s–I 5p) and nonbonding states, as determined by projected crystal orbital Hamilton population (pCOHP) analysis (Fig. 6.6). The conduction band minimum of Cs$_2$SnI$_6$ is comprised of antibonding I 5p and Sn 5s states. In contrast, the lowest unoccupied band of Cs$_2$TeI$_6$ is formed of antibonding I 5p and Te 5p states. From analysis of the band structure and pCOHP data, it is apparent that the band forming the lower conduction band in Cs$_2$SnI$_6$ becomes formally occupied upon moving to Cs$_2$TeI$_6$ due to the presence of the 2 extra electrons provided by Te. The next highest states above the conduction band minimum in Cs$_2$SnI$_6$ (found at around 4.5 eV) appear to be shifted down in energy to form the lower conduction band in Cs$_2$TeI$_6$.

6.3.3 Origin of Occupied Sn s Electrons

As observed in other hybrid DFT work on Cs_2SnI_6, we note the presence of occupied Sn $5s$ states $\sim 7\,eV$ beneath the valence band edge (Fig. 6.5a) [21, 31]. The existence of these states, despite the presence of Sn in the 4+ oxidation state, has been the topic of much interest in the literature [31, 39]. Indeed, based on the electronic configuration of Sn^{4+} ($s^0 p^0$), one might expect to only see *unoccupied* s and p contributions in the conduction band.

Comparison to the analogous three-dimensional perovskite, $CsSnI_3$, provides insight into the origin of this behaviour. Moving from $CsSnI_3$ to Cs_2SnI_6, the introduction of ordered Sn vacancies results in the effective addition of 2 holes per formula unit. Recent ^{119}Sn-Mössbauer spectroscopy and K-edge X-ray spectroscopy measurements, reveal the Sn $5s$ charge distribution is reduced going from $CsSnI_3$ to Cs_2SnI_6, as expected based on a formal oxidation state model [39]. The loss in the Sn $5s$ electrons, however, is accompanied by an increase in the charge density of the Sn $5p$ electrons, thereby acting to replenish most of the charge lost. This observation is confirmed by hybrid DFT calculations, which show that the transformation from $CsSnI_3$ to Cs_2SnI_6, results in a loss of 0.6 Sn s electrons but concomitant gain of 0.2 p electrons [39]. Accordingly, the total change in the physical charge surrounding the Sn atom is just 0.4 electrons, significantly less than expected. It has previously been suggested that the observation of occupied Sn s states arises from the presence of Sn in the +2 oxidation state [31]. The calculations described above, however, dispute this picture, proving instead that the Sn s orbitals are half occupied. As such, Dalpian et al. have proposed this behaviour can be explained through a *self-regulating response* mechanism [39, 40].

This mechanism can be understood by considering a simplified orbital bonding picture for both compounds (Fig. 6.7). The covalent bonding in the SnI_6 octahedral complex is dictated by symmetry constraints: specifically, the Sn possesses spherical s states and dumbbell p states, whereas the I_6 cage (O_h point group) shows p orbitals split into several symmetry adapted linear combinations. The Sn $5s$ electrons can only hybridise with I $5p$ A_{1g} states, while the Sn $5p$ states must couple with the I $5p$ T_{1u} states. The remaining I p states, with E_g, T_{1g}, T_{2g}, and T_{2u} symmetries have no compatible Sn states to bond with and will therefore be nonbonding.

In $CsSnI_3$, both the Sn $5s$–I $5p$ A_{1g} bonding and antibonding orbitals are occupied, resulting in a fully occupied Sn $5s$ shell (Fig. 6.7a). In contrast, the Sn $5p$–I $5p$ T_{1u} bonding orbital is occupied, whereas the antibonding is not, resulting in a band gap transition from the antibonding A_{1g} states to the antibonding T_{1u} states. Moving from $CsSnI_3$ to Cs_2SnI_6, the introduction of two holes (due to the regular Sn-site vacancies), causes the Fermi level to shift down, thereby depopulating the antibonding A_{1g} states (Fig. 6.7b). As these states contain greater I $5p$ than Sn $5p$ character, the holes are disproportionately distributed over the I_6 octahedra, accounting for the smaller than expected 0.6 electron loss in the Sn s density. The transition from $CsSnI_3$ to Cs_2SnI_6, is accompanied by a 8% reduction in the bond lengths of the SnI_6 octahedra, resulting in increased Sn-I orbital coupling [39]. This acts to push

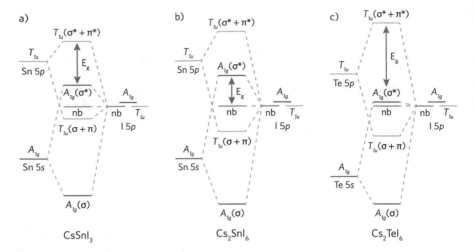

Fig. 6.7 Simplified molecular orbital picture of **a** SnI_6 in $CsSnI_3$, **b** SnI_6 in Cs_2SnI_6, and **c** TeI_6 in Cs_2TeI_6. A_{1g}, T_{1u}, and nonbonding (nb) orbitals indicated by blue, orange, and green lines, respectively

the bonding Sn $5p$–I $5p$ T_{1u} states lower in energy, while simultaneously increasing their Sn $5p$ character. Accordingly, the greater hybridisation results in an increase of occupied Sn $5p$ states, providing an explanation for the greater Sn p density seen in experiment. The remaining holes are instead localised in a complex arrangement throughout the iodine close-packed lattice, as indicated by charge density isosurfaces [39]. This subtle feedback loop—namely, greater occupation of the bonding Sn $5p$ orbitals upon depopulation of the antibonding Sn $5s$ states—effectively acts to minimise the effects of charge perturbation caused by vacancy formation and has been termed a self-regulating response [39, 40].

The molecular orbital picture can be extended to understand the behaviour of Cs_2TeI_6. The $5s^2 p^0$ electron configuration of Te^{4+}, effectively adds two electrons per formula unit, thereby repopulating the A_{1g} antibonding orbitals depopulated in the transition from $CsSnI_3$ to Cs_2SnI_6 (Fig. 6.7c). Due to the higher binding energy of the Te $5s$ orbitals (in comparison to the Sn $5s$), these states occur further away in energy from the I $5p$ states. This results in a significant reduction in hybridisation, causing the antibonding A_{1g} orbitals to sit very close in energy to the nonbonding I $5p$ states. This is confirmed by the orbital projected band structures, which show the majority of the Te $5s$ states localised around $12\,eV$ beneath the valence band maximum, with only a small number of states present at the band edge. In contrast, the larger binding energy of the Te $5p$ states brings them closer in energy to the I $5p$, increasing the hybridisation and raising the antibonding T_{1u} states. Combined, this behaviour has the effect of significantly raising the band gap of Cs_2TeI_6 *versus* Cs_2SnI_6.

6.3.4 Nature of the Cs₂SnI₆ Band Gap

In order to investigate the difference between the direct fundamental band gap and the experimentally observed optical band gap of Cs_2SnI_6, we have calculated the transition matrix elements for the transitions between the valence and conduction band states, using HSE06+SOC. The transitions were categorised as dipole allowed or disallowed, based on the magnitude of the matrix element square ($|M|^2$). Specifically, a transition was deemed allowed if $|M|^2 > 10^{-3}$eV^{-2} Å^{-2}, else it was considered dipole forbidden, as described in Ref. [41]. Analysis of the direct valence band to conduction band transitions, indicates that photo-excitation from the twofold-degenerate valence band maximum to the conduction band minimum at Γ (A_{1g} symmetry), is dipole forbidden. This effect is well known in materials whose crystal structures possess a centre of inversion [42, 43] and results as strong optical transitions are only permitted between states of opposing parity. Accordingly, the nonbonding states I $5p$ states at the top of the valence band must possess either E_g, T_{1g}, or T_{2g} symmetry. Transitions from the doubly degenerate bands 0.02 eV beneath the valence band maximum were also found to be dipole forbidden. It is only from 0.35 eV below the valence band maximum that strong transitions are observed, as illustrated in Fig. 6.8. These bands must therefore be the nonbonding I $5p$ states with T_{2u} symmetry. As such, the onset of optical absorption occurs at 1.32 eV, considerably larger than the direct band gap of 0.97 eV and in much better agreement with the experimental optical band gap of 1.26 eV. Accordingly, we believe that previous studies have erroneously employed large amounts of HF exchange (α parameter in the HSE functional) in order to mistakenly fit the fundamental band gap to the experimental optical band gap [19, 21, 31].

The symmetry forbidden fundamental band gap provides a possible explanation for the non-linear trend in optical gaps seen across the solid-solution series, $Cs_2Sn_{1-x}Te_xI_6$. As demonstrated in experiments performed by our collaborators, Maughan et al. [1] tellurium incorporation only results in a slight increase in the

Fig. 6.8 Band structure of Cs₂SnI₆, calculated using HSE06+SOC. The bands highlighted in green, indicate the states involved in the allowed optical band gap at Γ. The valence band maximum is set to 0 eV

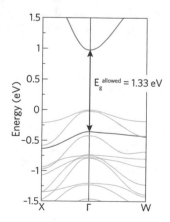

optical band gap between x = 0 and 0.75, after which a steep rise is observed (see
Fig. 6.3d). It is possible that the introduction of small amounts of Te will break
the symmetry of the frontier bonding states, causing the fundamental band gap to
become weakly dipole allowed. In this way, the rise in the band gap expected upon Te
substitution is offset by greater absorption occurring from the band edges. When the
fundamental band gap becomes fully allowed, a sharper increase in the absorption
will be observed, as now the band gap is solely dictated by the rise in the antibonding
T_{1u} states and lowering of the antibonding A_{1g} states (as discussed in the previous
section).

6.3.5   Effective Masses

The effective masses of Cs_2SnI_6 and Cs_2TeI_6 are summarised in Table 6.2. The
electron effective masses (m_e^*) of Cs_2SnI_6 were found to be relatively small, with
masses of $0.33m_0$ seen for both the $\Gamma \rightarrow X$ (along [1 0 0]) and $\Gamma \rightarrow L$ (along [0.5
0.5 0.5]) directions. The effective masses of Cs_2TeI_6 show greater anisotropy, with
masses of $0.45m_0$ and $0.30m_0$ seen for the $\Gamma \rightarrow L$ and $W \rightarrow L$ (along [0 0.5 −0.5])
directions, respectively. We note, that while there are slight differences in the effective
masses between the two compounds, these are insufficient to account for the dramatic
reduction in carrier mobilities seen in experiment when moving from Cs_2SnI_6 to
Cs_2TeI_6, as discussed in Sect. 6.1.1. The masses are slightly larger than those seen
in the three-dimensional hybrid perovskites, $CH_3NH_3PbI_3$ and $CH_3NH_3SnI_3$, which
show electron effective masses of $0.15m_0$ and $0.28m_0$, respectively [12, 44].

In contrast to the hybrid perovskites, in which the hole effective masses (m_h^*)
are smaller than the electron effective masses $(m_h^* = 0.12m_0$ and $0.13m_0$, for
$CH_3NH_3PbI_3$ and $CH_3NH_3SnI_3$), [12, 44] the vacancy-ordered perovskites do not
show a similar trend. In Cs_2SnI_6, the hole effective masses are reasonably large,
around $0.95m_0$ for both directions. Cs_2TeI_6, again shows greater anisotropy, with
hole effective masses of $0.65m_0$ and $3.00m_0$ seen for the $\Gamma \rightarrow X$ and $W \rightarrow L$ directions. In both cases, this can be attributed to a valence band dominated by non-bonding
I $5p$ states, resulting in limited dispersion.

Table 6.2 Electron (m_e^*) and hole (m_h^*) effective masses of Cs_2SnI_6 and Cs_2TeI_6, calculated using
HSE06+SOC. Masses given in units of electron rest mass, m_0

	m_e^*			m_h^*		
Compound	$\Gamma \rightarrow X$	$\Gamma \rightarrow L$	$W \rightarrow L$	$\Gamma \rightarrow X$	$\Gamma \rightarrow L$	$W \rightarrow L$
Cs_2SnI_6	0.33	0.33	–	0.96	0.93	–
Cs_2TeI_6	–	0.45	0.30	0.65	–	3.00

6.3.6 Reconciling Small Effective Masses and Ordered-Vacancies

The presence of dispersive conduction band states containing significant Sn $5s$ contributions despite the existence of regular Sn vacancies, is, at first glance, puzzling. To investigate this behaviour, we have calculated charge density isosurfaces of the lowest conduction band of (Fig. 6.9), at the Γ and X points in the Brillouin zone. At the Γ point, the A_{1g} antibonding orbital is in phase with the A_{1g} orbital in the adjacent cell (Fig. 6.9b), leading to a bonding interaction between neighbouring octahedra (Fig. 6.9c). This interaction occurs through significant portions of empty space, due to the disperse projection of the iodine $5p$ orbitals into the interoctahedral void and acts to stabilise the bonding at this $\mathbf{k}$-point. In contrast, at the X point, the A_{1g} antibonding orbital is out of phase with the A_{1g} orbital in the adjacent cell (Fig. 6.9d), resulting in an antibonding interaction between neighbouring octahedra (Fig. 6.9e). As such, the energy of the interaction at the X point is significantly higher, giving rise to reasonably high band dispersion.

Interestingly, the bonding counterpart of the A_{1g} orbital (highlighted in 6.10a) shows significantly less dispersion. Again, this can be understood through charge density isosurfaces of the band at the Γ and X points. At Γ, there is a similar bonding interaction between octahedra in neighbouring cells (Fig. 6.10b), however, there is significantly less I $5p$ projection into the void region (Fig. 6.10c). The interaction at X is again more antibonding (Fig. 6.10d), however, due to the reduced I p contribution in the interoctahedral void (Fig. 6.10e), the destabilisation effect is dramatically reduced. Both effects combine to produce a band with limited dispersion across

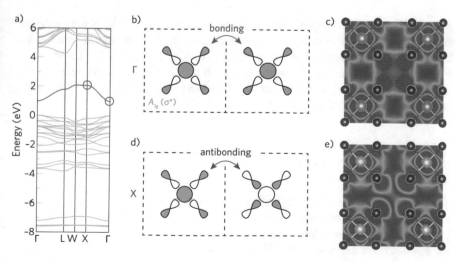

Fig. 6.9 **a** Band structure of Cs_2SnI_6, highlighting the A_{1g} antibonding states. Simplified bonding diagram (**a** and **d**) and charge density isosurfaces (**c** and **e**) of the highlighted band at the Γ and X points. The isosurface level was set to 0.008 eV Å^{-3}

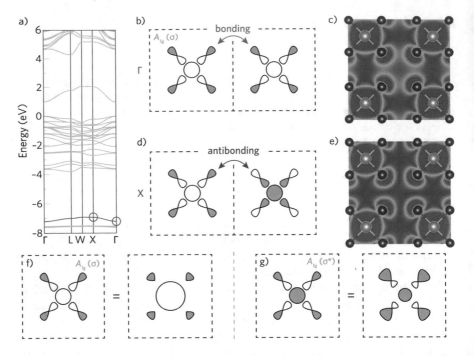

Fig. 6.10 a Band structure of Cs_2SnI_6, highlighting the A_{1g} bonding states. Simplified bonding diagram (**a** and **d**) and charge density isosurfaces **c, e** of the highlighted band at the Γ and X points. The isosurface level was set to 0.008 eV Å^{-3}. Schematic of electron distribution in a **f** bonding and **g** antibonding A_{1g} orbital

k-space. The reduced I p projection into the void region can be attributed to several factors: primarily, the bonding A_{1g} orbital contains less I p character than the anti-bonding state, resulting in less overall charge density present on the iodine atoms. Furthermore, the bonding A_{1g} molecular orbital intrinsically shows reduced electron density projecting outside the octahedra when compared to the A_{1g} antibonding orbital, as depicted in Fig. 6.10f, g.

6.3.7 Band Alignments

The band alignments of Cs_2SnI_6 and Cs_2TeI_6 relative to the vacuum level, calculated using HSE06 with a correction for spin-orbit coupling, are shown in Fig. 6.11. As predicted based on the simplified molecular orbital picture presented in Fig. 6.7, the valence band maximum, dominated by nonbonding I $5p$ orbitals, is effectively pinned in energy across both compounds. The calculated ionisation potential of $\sim$5.80 eV is in good agreement with the experimental value of $\sim$5.92 eV for Cs_2SnI_6 determined by X-ray photoelectron spectroscopy [1]. The ionisation potential of

Fig. 6.11 Experimental (exp.) and calculated (HSE06+SOC) band alignment of Cs_2SnI_6 and Cs_2TeI_6 relative to the vacuum level. The experimental ionisation potential of Cs_2SnI_6 was determined by X-ray photoelectron spectroscopy, with the electron affinity calculated based on the HSE06+SOC band gap

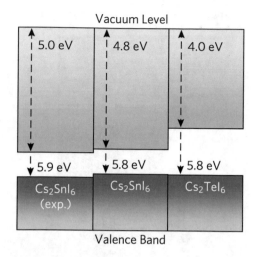

Cs_2SnI_6 is greater than that of its three-dimensional counter part, $CsSnI_3$ (5.74 eV), [33] as expected due to the population of the antibonding A_{1g} states at the top of the valence band. Due to the fixed ionisation potential, the electron affinity is directly controlled by the energy of the conduction band minimum. In Cs_2SnI_6 and Cs_2TeI_6, the antibonding Sn s–I p and antibonding Te p–I p states give rise to electron affinities of 4.8 eV and 4.0 eV, respectively.

6.3.8 Defect Chemistry

The experimental observation of a dramatic reduction in conductivity upon tellurium substitution requires an investigation into the defect properties of Cs_2SnI_6 and Cs_2TeI_6. A plot of the experimentally accessible chemical potentials for both compounds across a (μ_{Cs}, μ_{Te}) plane [45, 46] is provided in Fig. 6.12. The stability field is limited by the host conditions ($2\mu_{Cs} + \mu_{Sn} = \Delta_f H^{Cs_2SnI_6}$ and $2\mu_{Cs} + \mu_{Te} = \Delta_f H^{Cs_2TeI_6}$) to give the limits of Cs/Sn rich, Cs poor, and Sn poor environments for Cs_2SnI_6, with analogous environments for Cs_2TeI_6. Taking into account the limits imposed by the competing binary and ternary phases results in the stable regions shaded in orange. The accessible range of chemical potential space is significantly larger for Cs_2TeI_6, suggesting that formation of Cs_2SnI_6 will require greater sensitivity to the stoichiometric quantities of starting materials. Two potential environments have been highlighted, termed A and B, which correspond to Cs/Sn-rich, I-poor (A) and Cs/Sn-poor, I-rich (B) conditions. For both compounds, we have explicitly considered the formation energies of the defects using the chemical potentials at the A point, as a representative middle ground. We note, however, that the B point is expected to most favour the formation of n-type V_I defects.

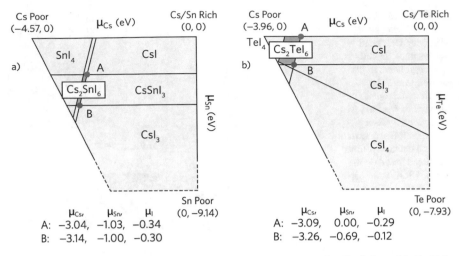

Fig. 6.12 Illustration of the accessible chemical potential ranges of **a** Cs_2SnI_6 and **b** Cs_2TeI_6. Constraints imposed by the formation of competing binary and ternary compounds result in the stable region indicated in orange

Previous research into $CsSnI_3$, has identified iodine vacancies (V_I) as low energy shallow defects that are likely to impact conductivity [47]. As such, to investigate the mechanism of conductivity in Cs_2SnI_6 and Cs_2TeI_6, we have considered the formation of donor V_I defects in both compounds. A plot of formation energy as function of Fermi level, under the chemical potentials at environment A, is provided in Fig. 6.13. For Cs_2SnI_6, the range of formation energies across all chemical potential environments is small (0.14–0.39 eV), indicating that these defects are likely to be present in high concentrations. Furthermore, V_I posses a relatively shallow $+1/0$ transition level 0.07 eV below the conduction band minimum, suggesting a source for the native n-type conductivity seen in experiment. Previous calculations have identified V_I as an ultra-deep donor, with a transition level 0.52 eV beneath the conduction band minimum [21]. Due the larger supercell size used in our calculations (232 vs. 72 atoms), [21] in addition to our correct treatment of the fundamental band gap—specifically, we do not erroneously adjust the amount of HF exchange to fit to the optical band gap—we believe our calculations are likely a better representation of the true defect chemistry. Indeed, our calculated transition level is more in keeping with a system displaying native n-type conductivity, with carrier concentrations up to $\sim 10^{17}$, as observed by our collaborators at Colorado State University [1].

Moving to Cs_2TeI_6, the formation energy of V_I, across all chemical potential environments, is considerably larger (0.56–0.84 eV), indicating there will be fewer iodine vacancy defects present. Additionally, the $+1/0$ transition state is found ultra-deep within the band gap (0.63 eV below the conduction band minimum). As such, V_I defects in Cs_2TeI_6 cannot produce samples with high n-type carrier concentrations. Interestingly, in both Cs_2SnI_6 and Cs_2TeI_6, the V_I $+1/0$ defect level appears to be fixed in energy, around 0.9 eV above the valence band maximum. These results

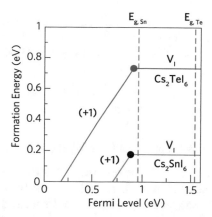

Fig. 6.13 HSE06+SOC calculated defect formation energies for iodine vacancies in Cs_2SnI_6 (red) and Cs_2TeI_6 (green), under the chemical potential environment A, identified in Fig. 6.12. Sloped lines indicate the +1 charge state and the solid dots represent the +1/0 transition levels. Dashed lines indicate the fundamental band gap of each material

suggest a possible source of the insulating behaviour observed in the solid-solution series, $Cs_2Sn_{1-x}Te_xI_6$ (Fig. 6.3). Namely, due to the pinned energy of the valence band maximum, the depth of the V_I +1/0 transition level will depend solely on the electron affinity of the system. Accordingly, as the system becomes more Te rich, the conduction band minimum will rise in energy, resulting in a reduction in n-type carrier concentrations and lower levels of conductivity.

6.4 Conclusions

In this chapter, we have investigated the fundamental optoelectronic properties of the vacancy-ordered defect perovskites Cs_2SnI_6 and Cs_2TeI_6, in collaboration with our experimental partners Maughan et al. Using first-principles relativistic electronic structure theory, we show that Cs_2SnI_6 and Cs_2TeI_6 possess direct and indirect fundamental band gaps, respectively. Furthermore, we reveal that the fundamental band gap of Cs_2SnI_6 is symmetry disallowed, thereby providing an explanation for the apparent mismatch between theory and experiment previously discussed in the literature. The observation of dispersive conduction band states containing significant Sn contributions, despite the presence of regular vacancies, is attributed to interoctahedral I–I interactions than span the voids in the lattice. We further reveal that these interaction are facilitated by the shape of the Sn $5s$–I $5p$ A_{1g} antibonding orbital, which shows greater electron density projecting outside the octahedra. The trend in effective masses is insufficient to account for the dramatic reduction in carrier concentrations and electron mobilities seen upon tellurium substitution for the solid-solution series $Cs_2Sn_{1-x}Te_xI_6$. Instead, we attribute this behaviour to the depth of iodine vacancy defects, which are shallowest for Cs_2SnI_6 and become deeper with increasing tellurium incorporation. These results provide an explanation for the puzzling experimental observables detailed by our experimental collaborators.

Notes

The work discussed in this chapter was featured in one publication:

1. Maughan AE, Ganose AM, Bordelon MM, Miller EM, Scanlon DO, Neilson JR (2016) Defect tolerance to intolerance in the vacancy ordered double perovskite semiconductors Cs_2SnI_6 and Cs_2TeI_6. J Amer Chem Soc 138:8453–8464

The optimised crystal structures, for all compounds discussed in this chapter, are provided in an online repository [32].

References

1. Maughan AE, Ganose AM, Bordelon MM, Miller EM, Scanlon DO, Neilson JR (2016) Defect tolerance to intolerance in the vacancy-ordered double perovskite semiconductors Cs_2SnI_6 and Cs_2TeI_6. J Am Chem Soc 138:8453–8464
2. Hao F, Stoumpos CC, Cao DH, Chang RPH, Kanatzidis MG (2014) Lead-free solid-state organic-inorganic halide perovskite solar cells. Nature Photon 8:489–494
3. Noel NK, Stranks SD, Abate A, Wehrenfennig C, Guarnera S, Haghighirad A, Sadhanala A, Eperon GE, Pathak SK, Johnston MB, Petrozza A, Herz L, Snaith H (2014) Lead-free organic-inorganic tin halide perovskites for photovoltaic applications. Energy Environ Sci 7:3061–3068
4. Chung I, Lee B, He J, Chang RP, Kanatzidis MG (2012) All-solid-state dye-sensitized solar cells with high efficiency. Nature 485:486–489
5. Sabba D, Mulmudi HK, Prabhakar RR, Krishnamoorthy T, Baikie T, Boix PP, Mhaisalkar S, Mathews N (2015) Impact of anionic Br^- substitution on open circuit voltage in lead free perovskite $(CsSnI_{3-x}Br_x)$ Solar Cells J Phys Chem C 119:1763–1767
6. Kumar MH, Dharani S, Leong WL, Boix PP, Prabhakar RR, Baikie T, Shi C, Ding H, Ramesh R, Asta M, Graetzel M, Mhaisalkar SG, Mathews N (2014) Lead-free halide perovskite solar cells with high photocurrents realized through vacancy modulation. Adv Mater 26:7122–7127
7. Shao S, Liu J, Portale G, Fang H-H, Blake GR, ten Brink GH, Koster LJA, Loi MA (2018) Highly reproducible Sn-based hybrid perovskite solar cells with 9% efficiency. Adv Energy Mater 8:1702019
8. Chung I, Song J-H, Im J, Androulakis J, Malliakas CD, Li H, Freeman AJ, Kenney JT, Kanatzidis MG (2012) $CsSnI_3$: semiconductor or metal? High electrical conductivity and strong near-infrared photoluminescence from a single material. High hole mobility and phase-transitions. J Am Chem Soc 134:8579–8587
9. Yu C, Ren Y, Chen Z, Shum K (2013) First-principles study of structural phase transitions in $CsSnI_3$. J Appl Phys 114:163505
10. da Silva EL, Skelton JM, Parker SC, Walsh A, Silva EL, Skelton JM, Parker SC, Walsh A (2015) Phase stability and transformations in the halide perovskite $CsSnI_3$. Phys Rev B 91:1–12
11. Brenner TM, Egger DA, Kronik L, Hodes G, Cahen D (2016) Hybrid organic-inorganic perovskites: low-cost semiconductors with intriguing charge-transport properties. Nat Rev Mater 1:15007

12. Umari P, Mosconi E, De Angelis F (2014) Relativistic GW calculations on $CH_3NH_3PbI_3$ and $CH_3NH_3SnI_3$ perovskites for solar cell applications. Sci Rep 4:4467

13. Walsh A (2015) Principles of chemical bonding and band gap engineering in hybrid organic-inorganic halide perovskites. J Mater Chem C 119:5755–5760

14. Anderson MT, Greenwood KB, Taylor GA, Poeppelmeier KR (1993) B-cation arrangements in double perovskites. Prog Solid State Chem 22:197–233

15. McClure ET, Ball MR, Windl W, Woodward PM (2016) Cs_2AgBiX_6 (X= Br, Cl): new visible light absorbing, lead-free halide perovskite semiconductors. Chem Mater 28:1348–1354

16. Slavney AH, Hu T, Lindenberg AM, Karunadasa HI (2016) A bismuth-halide double perovskite with long carrier recombination lifetime for photovoltaic applications. J Am Chem Soc 138:2138–2141

17. Savory CN, Walsh A, Scanlon DO (2016) Can Pb-free halide double perovskites support high-efficiency solar cells? ACS Energy Lett 1:949–955

18. Lee B, Stoumpos CC, Zhou N, Hao F, Malliakas C, Yeh C-Y, Marks TJ, Kanatzidis MG, Chang RP (2014) Air-stable molecular semiconducting iodosalts for solar cell applications: Cs_2SnI_6 as a hole conductor. J Am Chem Soc 136:15379–15385

19. Saparov B, Sun J-P, Meng W, Xiao Z, Duan H-S, Gunawan O, Shin D, Hill IG, Yan Y, Mitzi DB (2016) Thin-film deposition and characterization of a Sn-deficient perovskite derivative Cs_2SnI_6. Chem Mater 28:2315–2322

20. Jiang Y, Zhang H, Qiu X, Cao B (2017) The air and thermal stabilities of lead-free perovskite variant Cs_2SnI_6 powder. Mater Lett 199:50–52

21. Xiao Z, Zhou Y, Hosono H, Kamiya T (2015) Intrinsic defects in a photovoltaic perovskite variant Cs_2SnI_6 Phys. Chem Chem Phys 17:18900–18903

22. Kaltzoglou A, Antoniadou M, Kontos AG, Stoumpos CC, Perganti D, Siranidi E, Raptis V, Trohidou K, Psycharis V, Kanatzidis MG, Falaras P (2016) Optical-vibrational properties of the Cs_2SnX_6 (X= Cl, Br, I) defect perovskites and hole-transport efficiency in dye-sensitized solar cells. J Phys Chem C 120:11777–11785

23. Qiu X, Cao B, Yuan S, Chen X, Qiu Z, Jiang Y, Ye Q, Wang H, Zeng H, Liu J, Kanatzidis M (2017) From unstable $CsSnI_3$ to air-stable Cs_2SnI_6: A lead-free perovskite solar cell light absorber with bandgap of 1.48 eV and high absorption coefficient. Sol Energy Mater Sol Cells 159:227–234

24. Qiu X, Jiang Y, Zhang H, Qiu Z, Yuan S, Wang P, Cao B (2016) Lead-free mesoscopic Cs_2SnI_6 perovskite solar cells using different nanostructured ZnO nanorods as electron transport layers. Phys Status Solidi RRL 10:587–591

25. Lee B, Krenselewski A, Baik SI, Seidman DN, Chang RP (2017) Solution processing of air-stable molecular semiconducting iodosalts, $Cs_2SnI_{6-x}Br_x$, for potential solar cell applications. Sustain. Energy Fuels 1:710–724

26. Dolzhnikov DS, Wang C, Xu Y, Kanatzidis MG, Weiss EA (2017) Ligand-Free, Quantum-Confined Cs_2SnI_6 Perovskite Nanocrystals. Chem Mater 29:7901–7907

27. Pulay P (1969) Ab initio calculation of force constants and equilibrium geometries in poly-atomic molecules: I. Theory Mol Phys 17:197–204

28. Dronskowski R, Blöchl PE (1993) Crystal Orbital Hamilton Populations (COHP): energy-resolved visualization of chemical bonding in solids based on density-functional calculations. J Phys Chem 97:8617–8624

29. Maintz S, Deringer VL, Tchougréeff AL, Dronskowski R (2013) Analytic projection from plane-wave and paw wavefunctions and application to chemical-bonding analysis in solids. J Comput Chem 34:2557–2567

30. Wei S-H, Zunger A (1998) Calculated natural band offsets of all II–VI and III–V semiconductors: chemical trends and the role of cation d orbitals. Appl Phys Lett 72:2011–2013

31. Xiao Z, Hosono H, Kamiya T (2015) Origin of carrier generation in photovoltaic perovskite variant Cs_2SnI_6. Bull Chem Soc Jpn 88:1250–1255

32. https://github.com/SMTG-UCL/CSI-CTI. Accessed 2018 March 14

33. Zhang J, Yu C, Wang L, Li Y, Ren Y, Shum K (2014) Energy barrier at the N719-dye/$CsSnI_3$ interface for photogenerated holes in dye-sensitized solar cells. Sci Rep 4:6954

34. Peresh EY, Zubaka O, Sidei V, Barchii I, Kun S, Kun A (2002) Preparation, stability regions, and properties of M_2TeI_6 (M= Rb, Cs, Tl) crystals. Inorg Mater 38:859–863
35. Ranfagni A, Mugnai D, Bacci M, Viliani G, Fontana M (1983) The optical properties of thallium-like impurities in alkali-halide crystals. Adv Phys 32:823–905
36. Blasse G, Dirksen G, Abriel W (1987) The influence of distortion of the Te (IV) coordination octahedron on its luminescence. Chem Phys Lett 136:460–464
37. Drummen P, Donker H, Smit W, Blasse G (1988) Jahn-Teller distortion in the excited state of tellurium (IV) in Cs_2MCl_6 (M= Zr, Sn). Chem Phys Lett 144:460–462
38. Donker H, Smit W, Blasse G (1989) On the luminescence of Te_{4+} in A_2ZrCl_6 (A= Cs, Rb) and A_2ZrCl_6 (A= Cs, Rb, K). J Phys Chem Solids 50:603–609
39. Dalpian GM, Liu Q, Stoumpos CC, Douvalis AP, Balasubramanian M, Kanatzidis MG, Zunger A (2017) Changes in charge density vs changes in formal oxidation states: The case of Sn halide perovskites and their ordered vacancy analogues. Phys Rev Mater 1:025401
40. Raebiger H, Lany S, Zunger A (2008) Charge self-regulation upon changing the oxidation state of transition metals in insulators. Nature 453:763
41. Yu L, Zunger A (2012) Identification of potential photovoltaic absorbers based on first-principles spectroscopic screening of materials. Phys Rev Lett 108:068701
42. Walsh A, Da Silva JL, Wei S-H, Körber C, Klein A, Piper L, DeMasi A, Smith KE, Panaccione G, Torelli P (2008) Nature of the band gap of In_2O_3 revealed by first-principles calculations and X-ray spectroscopy. Phys Rev Lett 100:167402
43. Godinho KG, Carey JJ, Morgan BJ, Scanlon DO, Watson GW (2010) Understanding conductivity in $SrCu_2O_2$: stability, geometry and electronic structure of intrinsic defects from first principles. J Mater Chem 20:1086–1096
44. Frost JM, Butler KT, Brivio F, Hendon CH, Van Schilfgaarde M, Walsh A (2014) Atomistic origins of high-performance in hybrid halide perovskite solar cells. Nano Lett 14:2584–2590
45. Persson C, Zhao Y-J, Lany S, Zunger A (2005) n-type doping of $CuInSe_2$ and $CuGaSe_2$. Phys Rev B 72:035211
46. Scanlon DO (2013) Defect engineering of $BaSnO_3$ for high-performance transparent conducting oxide applications. Phys Rev B 87:161201
47. Xu P, Chen S, Xiang H-J, Gong X-G, Wei S-H (2014) Influence of defects and synthesis conditions on the photovoltaic performance of perovskite semiconductor $CsSnI_3$. Chem Mater 26:6068–6072

Part III
Bismuth-Based Absorbers

Chapter 7
Review: Bismuth-Based Photovoltaics

Bismuth-based solar absorbers have recently attracted attention as a result of the similarities between the bismuth halides and lead hybrid perovskites [1–3]. Both bismuth and lead adopt oxidation states two lower than their group valence, resulting in a stable $d^{10}s^2p^0$ electronic configuration [4]. Furthermore, the Bi^{3+} and Pb^{2+} ions form a large variety of compounds with rich structural diversity [3, 5–8]. Bismuth also experiences large relativistic effects that act to beneficially increase the conduction band width and stabilise the material with respect to oxidation [9–11]. Bismuth further presents an advantage over lead in that it is non-toxic and non-bioaccumulating, limiting the potential environmental impact if device encapsulation is compromised [12, 13]. The recent rise of bismuth in "green" catalysis and synthesis [14, 15] has prompted some to question whether bismuth absorbers can achieve similar success in photovoltaics [16].

7.1 Bismuth Sulfide

Bismuth sulfide (Bi_2S_3) possesses an optical band gap between 1.3 and 1.6 eV [17–19], strong optical absorption, and low toxicity, making it an attractive candidate for photovoltaics (Fig. 7.1). While bismuth sulfide was originally tested for use in photoelectrochemical cells in the 1980s [20, 21], it was not until recently that Bi_2S_3 was employed as a solar absorber. In particular, chemically deposited thin films of Bi_2S_3 were combined with lead chalcogenides, enabling devices with efficiencies of 0.5 % in 2011 [17] and 2.5 % in 2013 [22]. As Bi_2S_3 possesses intrinsic n-type conductivity [23], it has also been successfully employed in heterojunctions with

A. Ganose, *Atomic-Scale Insights into Emergent Photovoltaic Absorbers*, Springer Theses, https://doi.org/10.1007/978-3-030-55708-9_7

Fig. 7.1 Band structure of
Bi_2S_3 calculated using
HSE06 with spin–orbit
coupling. Adapted with
permission from Ref. [23].
Copyright (2016) American
Chemical Society

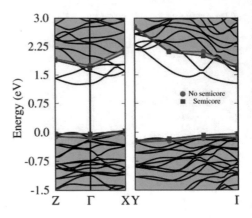

p-type materials such as PbS quantum dots [24, 25]. These devices have achieved almost 5 % power conversion efficiencies based on a "bulk–nano" heterojunction architecture. More recently, hybrid devices containing Bi_2S_3 along with the organic absorber P3HT have shown promising improvements in their efficiencies, reaching 3.3 % in 2015 [26–28]. DFT calculations on the defect properties of Bi_2S_3, however, suggest that performance may be limited due to the presence of deep trap states [29].

7.2 Bismuth Iodide

Bismuth iodide was recently identified—along with bismuth sulfide—as a potential solar absorber due to its suitable band gap and the possibility of high defect tolerance [1]. BiI_3 crystallises in a layered rhombohedral crystal structure [30] with a slightly indirect band gap of 1.67 eV (Fig. 7.2) [31, 32]. Initial devices produced by Lehner et al. obtained efficiencies of 0.3 % with the V_{oc} limited by poor alignment with the electron and hole contact layers [2, 33]. Subsequent work by Brandt et al. found that BiI_3 possessed strong optical absorption, despite the indirect band gap, but short carrier lifetimes of 180–240 ps indicating high carrier recombination [3]. In 2017, Hamdeh et al. reported devices with efficiencies up to 1.0 %, achieved through a solvent annealing process that optimised grain size and orientation [34]. More recent DFT work investigating the defect properties of BiI_3 reveals that all dominant point defects possess deep-level traps that can act as non-radiative recombination centres [35]. As such, it will be difficult for BiI_3 to match the performance of other emerging absorbers.

Fig. 7.2 a Crystal structure
and **b** band structure of BiI$_3$,
calculated using HSE06 with
spin–orbit coupling. Bismuth
and iodine atoms shown by
grey and purple spheres,
respectively. Reprinted from
Ref. [33], with the
permission of AIP
Publishing

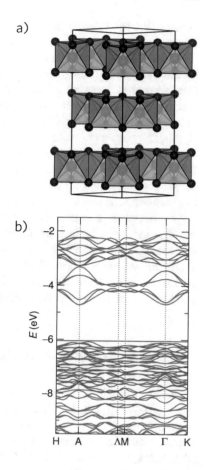

a)

b)

7.3 Silver Bismuth Sulfide and Iodide

Nanocrystalline silver bismuth sulfide has recently gained attention as a photovoltaic
absorber. Crystallising in a three-dimensional cubic distorted rocksalt structure [36,
37], bulk AgBiS$_2$ possesses an optical band gap of 0.9 eV [38]. Strong quantum con-
finement effects, however, allow for quantum dot thin-films with band gaps ranging
from 1.0–1.3 eV, depending on the size of crystallites, with strong optical absorption
above 10^5 cm^{-1} [38–40]. Initial devices containing AgBiS$_2$ quantum dots as a sensi-
tiser for TiO$_2$ demonstrated poor efficiencies of less than a percent due to poor fill factors and short-circuit current densities [37, 41]. More recently, devices by Bernechea
et al. containing tetramethylammonium iodide-treated AgBiS$_2$ nanocrystals in a *p–
i–n* solar cell architecture obtained record efficiencies of 6.3 % [42]. This performance was enabled by a combination of high short-circuit current (22 mA cm^{-2})
and fill factor (63 %), making AgBiS$_2$ competitive with the best bismuth and antimony solar absorbers. Performance was limited by a relatively small open-circuit

voltage (0.45 eV) due to high levels of trap-assisted recombination. Accordingly, improvements in thin film synthesis is expected to further improve efficiencies.

$AgBi_2I_7$ is an emerging absorber that has only been sparsely characterised since its discovery in 1979 [43]. Based on an investigation into the AgI–BiI_3 solid solution, Kim et al. reported a direct band gap of 1.86 eV for a crystal thought to be $AgBi_2I_7$ [44]. A subsequent investigation using a combined theoretical and experimental approach indicated the band gap and X-ray diffraction data was instead a significantly better match for an Ag-deficient $AgBiI_4$ type structure [45]. Regardless, when employed in a device with mesoporous TiO_2 as the electron contact layer and P3HT as the hole contact layer, an efficiency of 1.12 % was obtained. While the device showed a high fill factor of 67 %, its performance was limited by relatively low short-circuit current and open-circuit voltage. In 2017, [46] examined a series of $Ag_aBi_bI_x$ (where $x = a + 3b$) Rudorffites structures, including Ag_3BiI_6, Ag_2BiI_5, $AgBiI_4$, and $AgBi_2I_7$. Despite all compounds possessing relatively large band gaps in the range 1.79–1.83 eV, their champion device containing Ag_3BiI_6 on mesoporous TiO_2 achieved efficiencies of 4.3 %. While very promising for a proof-of-concept device, similar to many bismuth-based photovoltaic absorbers, additional attention will be required to achieve significant performance improvements in the further.

7.4   Caesium and Hybrid Bismuth Iodides

Caesium bismuth iodide ($Cs_3Bi_2I_9$) and its methylammonium analogue (($CH_3NH_3)_3Bi_2I_9$) have attracted attention due to the focus on their antimony counterparts [47]. At room temperature, both compounds crystallise in a "zero-dimensional" structure containing isolated face-sharing octahedral $[Bi_2I_9]^{3-}$ dimers (Fig. 7.3a). The Cs+ and $(CH_3NH_3)^+$ cations are non-bonding and serve to charge balance the structure [48–50]. Experimental work combined with density functional theory results, indicate an indirect band gap of 1.9 eV with the low dimensionality producing localised electronic states (Fig. 7.3b), suggesting carrier mobilities will be low [2]. The limited dispersion, however, gives rise to strong optical absorption, which may make these materials suitable for photovoltaics [51, 52].

Initial devices containing mesoporous TiO_2 with $Cs_3Bi_2I_9$ and $(CH_3NH_3)_3Bi_2I_9$ performed poorly, achieving efficiencies of ~1 % and 0.2 %, respectively [53]. In both cases, performance was limited by small short-circuit current and fill factor, with the open-circuit voltage remaining fairly large (~0.85 eV). Subsequent studies on $(CH_3NH_3)_3Bi_2I_9$ employing alternative hole contact layers showed small improvements in J_{sc} [54] and fill factor [55], but again suffered from low efficiencies. The photoluminescence behaviour of $(CH_3NH_3)_3Bi_2I_9$ indicated a charge-carrier recombination lifetime of 760 ps, a factor of 10^3 lower than those found in the best MAPI films [51]. Based on a low photoluminescence quantum efficiency of 0.4 % and the observation of defect states through X-ray photoelectron spectroscopy, non-radiative recombination pathways are thought to be the source of the small carrier lifetimes [53]. In 2018, use of vapour-assisted solution process deposition enabled

Fig. 7.3 **a** Crystal structure
and **b** band structure of
$Rb_3Bi_2I_9$ and $Cs_3Bi_2I_9$,
calculated using HSE06 with
spin–orbit coupling. Bismuth
and iodine atoms shown by
grey and purple spheres,
respectively. Adapted with
permission from Ref. [2].
Copyright (2015) American
Chemical Society

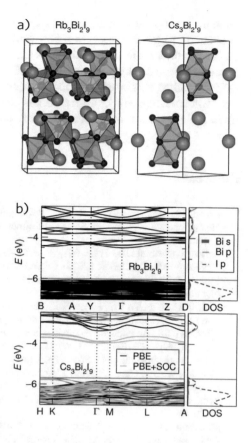

$(CH_3NH_3)_3Bi_2I_9$ based devices with efficiencies up to 3.17 %. Impressively, the devices showed very high open-circuit voltages up to 1.0 eV and less than 0.1 % loss in efficiency when exposed to air for 60 days. Such a high stability is a strong motivator for additional work on improving device efficiencies.

Another hybrid inorganic-organic bismuth iodide that is of interest as a solar absorber is the one-dimensional hexanediammonium bismuth iodide, $(H_3NC_6H_12NH_3)BiI_5$. Possessing a relatively large band gap around 2 eV, it was recently employed in a device with mesoporous TiO_2 [56]. Despite excellent film coverage and thermal stability, the device performed poorly with a low open-circuit voltage and short-circuit of 0.38 eV and 0.10 mA cm^{-2}, respectively.

7.5   Outlook

The rise of the hybrid perovskites has stimulated research into lead-free alternatives. Bismuth-based materials are attractive due to their earth-abundant and non-toxic nature, combined with promising optoelectronic properties. Accordingly, the field of

bismuth-based photovoltaics has blossomed over the last 3–4 years and is expected
to continue to grow. Indeed, the best performing devices show comparable efficiencies to the best-in-class antimony- and tin-based alternatives. In particular, the
three-dimensional cubic structure of $AgBiS_2$ draws many parallels to the hybrid perovskites, and impressive efficiencies of over 6 % highlight its potential as an emerging
photovoltaic. Despite this, further work is clearly needed if bismuth-based materials
are to achieve success in an already highly competitive solar panel market.

References

1. Brandt RE, Stevanović V, Ginley DS, Buonassisi T (2015) Identifying defect-tolerant semiconductors with high minority-carrier lifetimes: beyond hybrid lead halide Perovskites. MRS
Commun 5:1–11
2. Lehner AJ, Fabini DH, Evans HA, Hébert C-A, Smock SR, Hu J, Wang H, Zwanziger JW,
Chabinyc ML, Seshadri R (2015) Crystal and Electronic Structures of Complex Bismuth
Iodides $A_3Bi_2I_9$ (A = K, Rb, Cs) related to perovskite: aiding the rational design of photovoltaics. Chem Mater 27:7137–7148
3. Brandt RE, Kurchin RC, Hoye RLZ, Poindexter JR, Wilson MWB, Sulekar S, Lenahan F,
Yen PXT, Stevanović V, Nino JC, Bawendi MG, Buonassisi T (2015) Investigation of Bismuth
triiodide (BiI_3) for photovoltaic applications. J Phys Chem Lett 6:4297–4302
4. Walsh A, Payne DJ, Egdell RG, Watson GW (2011) Stereochemistry of post-transition metal
oxides: revision of the classical lone pair model. Chem Soc Rev 40:4455–4463
5. Wu LM, Wu XT, Chen L (2009) Structural overview and structure-property relationships of
Iodoplumbate and Iodobismuthate. Coord Chem Rev 253:2787–2804
6. Mitzi DB, Chondroudis K, Kagan CR (2001) Organic-inorganic electronics. IBM J Res Dev
45:29–45
7. Demartin F, Gramaccioli C, Campostrini I (2010) Demicheleite-(I), BiSI, a new mineral from
La Fossa Crater, Vulcano, Aeolian Islands, Italy. Mineral Mag 74:141–145
8. Frit B, Mercurio J (1992) The crystal chemistry and dielectric properties of the Aurivillius
family of complex Bismuth oxides with perovskite-like layered structures. J Alloys Compd
188:27–35
9. Umari P, Mosconi E, De Angelis F (2014) Relativistic GW calculations on $CH_3NH_3PbI_3$ and
$CH_3NH_3SnI_3$ perovskites for solar cell applications. Sci Rep 4:4467
10. Ganose AM, Savory CN, Scanlon DO (2017) Beyond methylammonium lead iodide: prospects
for the emergent field of ns^2 containing solar absorbers. Chem Commun 53:20–44
11. Brivio F, Butler KT, Walsh A, Van Schilfgaarde M (2014) Relativistic quasiparticle self-
consistent electronic structure of hybrid halide perovskite photovoltaic Absorbers. Phys Rev
B 89:155204
12. Slikkerveer A, de Wolff FA (1989) Pharmacokinetics and toxicity of Bismuth compounds. Med
Toxicol Adverse Drug Exp 4:303–323
13. Serfontein W, Mekel R (1979) Bismuth toxicity in man II. Review of Bismuth blood and urine
levels in patients after administration of therapeutic bismuth formulations in relation to the
problem of Bismuth toxicity in man. Res Commun Chem Pathol Pharmacol 26:391–411
14. Leonard NM, Wieland LC, Mohan RS (2002) Applications of Bismuth(III) compounds in
organic synthesis. Tetrahedron 58:8373–8397
15. Mohan R (2010) Green Bismuth. Nat Chem 2:336–336
16. Fabini DH, Labram JG, Lehner AJ, Bechtel JS, Evans HA, Van der Ven A, Wudl F, Chabinyc
ML, Seshadri R (2016) Main-group halide semiconductors derived from perovskite: distin-
guishing chemical, structural, and electronic aspects. Inorg Chem 56:11–25

17. Moreno-García H, Nair MTS, Nair PK (2011) Chemically deposited lead sulfide and bismuth sulfide thin films and Bi_2S_3/PbS solar cells. Thin Solid Films 519:2287–2295
18. ten Haaf S, Sträter H, Brüggemann R, Bauer GH, Felser C, Jakob G (2013) Physical vapor deposition of Bi_2S_3 as absorber material in thin film photovoltaics. Thin Solid Films 535:394–397
19. Filip MR, Patrick CE, Giustino F (2013) GW quasiparticle band structures of stibnite, antimonselite, bismuthinite, and guanajuatite. Phys Rev B 87:205125
20. Pramanik P, Bhattacharya RN (1980) A chemical method for deposition of thin film of Bi_2S_3. J Electrochem Soc 127:2087
21. Bhattacharya RN, Pramanik P (1982) Semiconductor liquid junction solar cell based on chemically deposited Bi_2S_3 thin film and some Semiconducting Properties of Bismuth Chalcogenides. J Electrochem Soc 129:332–335
22. Calixto-Rodriguez M, García HM, Nair MTS, Nair PK (2013) Antimony chalcogenide/lead selenide thin film solar cell with 2.5. ECS J Solid State Sci Technol 2:Q69–Q73
23. Tumelero MA, Faccio R, Pasa AA (2016) Unraveling the native conduction of trichalcogenides and its ideal band alignment for new photovoltaic Interfaces. J Phys Chem C 120:1390–1399
24. Rath AK, Bernechea M, Martinez L, Konstantatos G (2011) Solution-processed heterojunction solar cells based on p-type PbS quantum dots and n-type Bi_2S_3 Nanocrystals. Adv Mater 23:3712–3717
25. Rath AK, Bernechea M, Martinez L, de Arquer FPG, Osmond J, Konstantatos G (2012) Solution-processed inorganic bulk nano-heterojunctions and their application to solar cells. Nat Photon 6:529–534
26. Martinez L, Bernechea M, de Arquer FPG, Konstantatos G (2011) Near IR-sensitive, non-toxic, polymer/nanocrystal solar cells employing Bi_2S_3 as the electron acceptor. Adv Energy Mater 1:1029–1035
27. Martinez L, Stavrinadis A, Higuchi S, Diedenhofen SL, Bernechea M, Tajima K, Konstantatos G (2013) Hybrid solution-processed bulk heterojunction solar cells based on Bismuth sulfide nanocrystals. Phys Chem Chem Phys 15:5482–5487
28. Whittaker-Brooks L, Gao J, Hailey AK, Thomas CR, Yao N, Loo Y-L (2015) Bi_2S_3 nanowire networks as electron acceptor layers in solution-processed hybrid solar cells. J Mater Chem C 3:2686–2692
29. Han D, Du M-H, Dai C-M, Sun D, Chen S (2017) Influence of defects and dopants on the photovoltaic performance of Bi_2S_3: first-principles insights. J Mater Chem A 5:6200–6210
30. Ruck M (1995) Darstellung und Kristallstruktur von fehlordnungfreiem Bismuttriiodid. Z Kristallogr 210:650–655
31. Podraza NJ, Qiu W, Hinojosa BB, Xu H, Motyka MA, Phillpot SR, Baciak JE, Trolier-McKinstry S, Nino JC (2013) Band gap and structure of single crystal BiI_3: resolving discrepancies in literature. J Appl Phys 114:033110
32. Yorikawa H, Muramatsu S (2008) Theoretical study of crystal and electronic structures of BiI_3. J Phys Condens Matter 20:325220
33. Lehner AJ, Wang H, Fabini DH, Liman CD, Hébert C-A, Perry EE, Wang M, Bazan GC, Chabinyc ML, Seshadri R (2015) Electronic structure and photovoltaic application of BiI_3. Appl Phys Lett 107:131109
34. Hamdeh UH, Nelson RD, Ryan BJ, Bhattacharjee U, Petrich JW, Panthani MG (2016) Solution-processed BiI_3 thin films for photovoltaic applications: improved carrier collection via solvent annealing. Chem Mater 28:6567–6574
35. Cho SB, Gazquez J, Huang X, Myung Y, Banerjee P, Mishra R (2018) Intrinsic point defects and intergrowths in layered Bismuth triiodide. Phys Rev Mater 2:064602
36. Geller S, Wernick JH (1959) Ternary semiconducting compounds with sodium chloride-like structure: $AgSbSe_2$, $AgSbTe_2$, $AgBiS_2$, $AgBiSe_2$. Acta Crystallogr 12:46–54
37. Huang P-C, Yang W-C, Lee M-W (2013) $AgBiS_2$ semiconductor-sensitized solar cells. J Phys Chem C 117:18308–18314
38. Pejova B, Grozdanov I, Nesheva D, Petrova A (2008) Size-dependent properties of sonochemically synthesized three-dimensional arrays of close-packed Semiconducting $AgBiS_2$ Quantum Dots. Chem Mater 20:2551–2565

39. Pejova B, Nesheva D, Aneva Z, Petrova A (2011) Photoconductivity and relaxation dynamics in sonochemically synthesized assemblies of AgBiS$_2$ Quantum Dots. J Phys Chem C 115:37–46
40. Chen C, Qiu X, Ji S, Jia C, Ye C (2013) The synthesis of monodispersed AgBiS$_2$ quantum dots with a giant dielectric constant. CrystEngComm 15:7644–7648
41. Zhou S, Yang J, Li W, Jiang Q, Luo Y, Zhang D, Zhou Z, Li X (2016) Preparation and photovoltaic properties of ternary AgBiS$_2$ quantum dots sensitized TiO$_2$ nanorods Photoanodes by Electrochemical Atomic Layer Deposition. J Electrochem Soc 163:D63–D67
42. Bernechea M, Miller NC, Xercavins G, So D, Stavrinadis A, Konstantatos G (2016) Solution-processed solar cells based on environmentally friendly AgBiS$_2$ nanocrystals. Nat Photon 10:521–525
43. Fourcroy P, Palazzi M, Rivet J, Flahaut J, Céolin R (1979) Etude du systeme AgIBiI$_3$. Mater Res Bull 14:325–328
44. Kim Y, Yang Z, Jain A, Voznyy O, Kim G-H, Liu M, Quan LN, García de Arquer FP, Comin R, Fan JZ, Sargent EH (2016) Pure cubic-phase hybrid iodobismuthates AgBi2I$_7$ for thin-film photovoltaics. Angew Chem Int Ed 55:9586
45. Xiao Z, Meng W, Mitzi DB, Yan Y (2016) Crystal structure of AgBi$_2$I$_7$ thin films. J Phys Chem Lett 7:3903–3907
46. Turkevych I, Kazaoui S, Ito E, Urano T, Yamada K, Tomiyasu H, Yamagishi H, Kondo M, Aramaki S (2017) Photovoltaic rudorffites: lead-free silver Bismuth halides alternative to hybrid lead halide perovskites. ChemSusChem 10:3754–3759
47. Saparov B, Hong F, Sun J-P, Duan H-S, Meng W, Cameron S, Hill IG, Yan Y, Mitzi DB (2015) Thin-film preparation and characterization of Cs$_3$Sb$_2$I$_9$: a lead-free layered perovskite Semiconductor. Chem Mater 27:5622–5632
48. Lindquist O (1968) Crystal structure of caesium bismuth iodide Cs$_3$Bi$_2$I$_9$. Acta Chem Scand 22:2943–2952
49. Chabot B, Parthé E (1978) Cs$_3$Sb$_2$I$_9$ and Cs$_3$Bi$_2$I$_9$ with the hexagonal Cs$_3$Cr$_2$Cl$_9$ structure type. Acta Crystallogr Sect B Struct Sci 34:645–648
50. Eckhardt K, Bon V, Getzschmann J, Grothe J, Wisser FM, Kaskel S (2016) Crystallographic insights into (CH$_3$NH$_3$)$_3$(Bi$_2$I$_9$): a new lead-free hybrid organic-inorganic material as a potential absorber for photovoltaics. Chem Commun 52:3058–3060
51. Hoye RLZ, Brandt RE, Osherov A, Stevanović V, Stranks SD, Wilson MWB, Kim H, Akey AJ, Perkins JD, Kurchin RC, Poindexter JR, Wang EN, Bawendi MG, Bulović V, Buonassisi T (2016) Methylammonium bismuth iodide as a lead-free, stable hybrid organic-inorganic solar absorber. Chem Eur J 22:2605–2610
52. Kawai T, Shimanuki S (1993) Optical studies of (CH$_3$NH$_3$)$_3$Bi$_2$I$_9$ single crystals. Phys Status Solidi B 177:K43–K45
53. Park B-W, Philippe B, Zhang X, Rensmo H, Boschloo G, Johansson EMJ (2015) Bismuth based hybrid perovskites A$_3$Bi$_2$I$_9$ (a: methylammonium or cesium) for solar cell Application. Adv Mater 27:6806–6813
54. Lyu M, Yun J-H, Cai M, Jiao Y, Bernhardt PV, Zhang M, Wang Q, Du A, Wang H, Liu G, Wang L (2016) Organic-inorganic bismuth (III)-based material: a lead-free, air-stable and solution-processable light-absorber beyond organolead perovskites. Nano Res 9:692–702
55. Öz S, Hebig J-C, Jung E, Singh T, Lepcha A, Olthof S, Jan F, Gao Y, German R, van Loosdrecht PH, Meerholz K, Kirchartz T, Mathur S (2016) Zero-dimensional (CH$_3$NH$_3$)$_3$Bi$_2$I$_9$ perovskite for optoelectronic applications. Sol Energy Mater Sol Cells 158:195–201
56. Fabian DM, Ardo S (2016) Hybrid organic-inorganic solar cells based on bismuth iodide and 1,6-hexanediammonium dication. J Mater Chem A 4:6837–6841

Chapter 8
Bismuth Chalcoiodides

8.1 Introduction

In[1] 2015, Brandt et al. identified two bismuth-based materials, BiSI and BiSeI, as potential solar absorbers [3]. Both compounds are isostructural, crystallising in a 1D ribbon structure held together by weak dispersive forces (Fig. 8.1) [4, 5]. As proposed by Zhou et al., this type of structure may have many advantages for solar applications. If the crystal is oriented so that the 1D ribbons are vertically aligned on the substrate (e.g. the hole or electron contact materials in a photovoltaic device), a direct pathway along the covalently bonded axis of the ribbons connects the device contacts, potentially allowing for efficient carrier extraction routes. Furthermore, if vertically aligned growth is enforced, any grain boundaries or disconnections within the crystal should not disrupt the conduction pathways and instead occur parallel to the ribbons, avoiding the formation of dangling bonds and associated defect trap sites [6]. This is particularly important for solar cells, where defect sites are a primary cause of recombination losses. The advantages of this geometry have recently been demonstrated in the similarly structured compound, Sb_2Se_3, where devices containing vertically oriented films achieved over 2% greater efficiency than their non-oriented counterparts [6].

The only functioning photovoltaic devices containing BiSI and BiSeI reported in the literature, were produced by Hahn et al. in 2012 [7, 8]. Both compounds were found to possess large absorption coefficients and band gaps of 1.57 eV and 1.29 eV, respectively, ideal for solar cell applications. Their devices—employing p-CuSCN as the hole transport layer and fluorine-doped tin oxide (FTO) as the transparent contact—performed poorly, only achieving efficiencies up to 0.012%. The authors attribute the unsatisfactory performance of their devices to small hole diffusion lengths of 50 nm, [8] proposing that high levels of charge screening at the

---

[1]Parts of this chapter have been reproduced with permission from Refs. [1, 2]—Published by The Royal Society of Chemistry.

A. Ganose, *Atomic-Scale Insights into Emergent Photovoltaic Absorbers*, Springer Theses, https://doi.org/10.1007/978-3-030-55708-9_8

Fig. 8.1 BiSI crystal
structure viewed along **a** the
[100] direction, **b** a
perspective highlighting the
1D nature of the ribbons, and
c the [001] direction. Se, Bi,
and I atoms are denoted by
green, grey, and purple
spheres, respectively

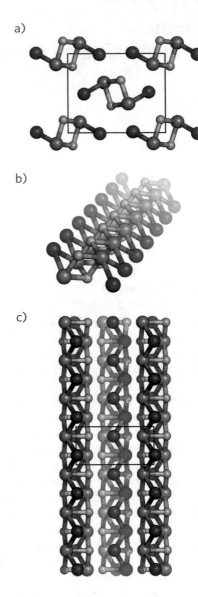

a)

b)

c)

heterojunction interface prevented effective electron–hole separation and resulted
in increased rates of recombination. This explanation is, however, at odds to other
solar absorbers with large dielectric constants, such as MAPI where efficient charge
screening is thought to provide increased tolerance to defects and *reduce* unwanted
carrier recombination [3, 9, 10]. Instead, based on our previous discussion of the 1D
ribbon structure, it is likely that the small hole diffusion lengths resulted from the
formation of polycrystalline films, which disrupt the pathways for carrier extraction
and increase recombination rates. More recently, advancements in low-temperature

synthesis routes has enabled BiSI films with improved photon-to-current conversion efficiency (IPCE) [11]. As such, these materials possess many of the ideal properties required in efficient photovoltaics and, therefore, warrant a detailed analysis of their electronic properties.

8.2 Methodology

Calculations were performed using the Vienna *Ab initio* Simulation Package. A **k**-point mesh of Γ-centred $3 \times 6 \times 2$ and plane wave cutoff of 400 eV was found to converge the 12 atom unit cells of BiSI and BiSeI to within 1 meV/atom. During geometry optimisations, the cutoff was increased to 520 eV to avoid errors resulting from Pulay stress [12]. The structures were deemed converged when the forces totalled less than 10 meV Å^{-1}.

Several functionals were used in this work: for geometry relaxations, PBEsol and PBE were employed, with and without the addition of Grimme's D3 dispersion correction. Electronic properties were calculated using HSE06 with the addition of spin–orbit coupling effects (HSE06+SOC). The Brillouin zone for the *Pnma* space group, indicating the high-symmetry points explored in the band structure, is provided in Fig. 8.2. Density functional perturbation theory (DFPT) was employed, in combination with the PBEsol functional, to calculate the ionic contribution to the dielectric constants, with a denser **k**-point mesh of Γ-centred $8 \times 14 \times 6$ required to achieve convergence. Projected crystal orbital Hamilton population (pCOHP) analysis was performed using the LOBSTER program, based on wavefunctions calculated using HSE06 [13, 14].

For band alignment calculations, the core-level alignment approach of Wei and Zunger was employed, [15] using a slab model with 20 Å of vacuum and a 20 Å thick slab. The slab was cleaved along the non-polar (110) surface, due to the absence of any dangling bonds. Due to the size of the model, which precluded the use of

Fig. 8.2 Brillouin zone of the *Pnma* space group. Coordinates of the high symmetry points used for the band structures and effective masses: $\Gamma = (0, 0, 0)$; $Y = (1/2, 0, 0)$; $X = (0, 1/2, 0)$; $Z = (0, 0, 1/2)$; $U = (0, 1/2, 1/2)$; $T = (1/2, 0, 1/2)$; $S = (1/2, 1/2, 0)$; $R = (0, 1/2, 1/2)$

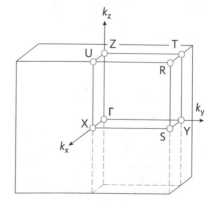

HSE06+SOC, band alignment calculations were performed using HSE06, with an explicit correction to the band gap and valence band maximum position taken from the HSE06+SOC calculated bulk.

Defect calculations were performed in a $2 \times 3 \times 1$ supercell containing 72 atoms, using the PBEsol functional and a Γ-centred $2 \times 3 \times 3$ **k**-point mesh. As PBEsol reproduces the experimental band gaps of both compounds, the defect transition levels will not need to be extrapolated to account for band gap underestimation. Despite this, as generalised-gradient approximation functionals can suffer from the self-interaction error, some unphysical delocalisation of defect states may occur. This may manifest as incorrect stabilisation of partially or unionised defects. The defect energies were corrected to account for use of a finite-sized supercell using the potential level alignment, band-filling, and image-charge corrections described in Chap. 3. The SC- FERMI code was utilised to calculate the self-consistent Fermi level and corresponding defect concentrations.

8.3 Results

8.3.1 Geometric Structure

BiSI and BiSeI share the same crystal structure as SbSI (*Pnma*), whereby 1D ribbons of BiChI are held together by weak van der Waals type interactions (Fig. 8.1). The ribbons are formed of distorted edge-sharing pseudo-octahedra, containing three Bi−Ch and two Bi−I bonds, with the lone pair on the Bi occupying a vacant site [4, 5]. The Bi − Ch bonds can be categorised into those occurring parallel (Bi−Ch$_{\parallel}$) or perpendicular (Bi−Ch$_{\perp}$) to the axis of the ribbons. Both structures were geometrical optimised using the PBE and PBEsol functionals, both with and without explicit treatment of dispersive interactions (Table 8.1). The PBEsol functional, in the absence of the D3 correction, was found to provide the most accurate description of lattice parameters and bond lengths, with all distances found to be within 1% of the experimental crystal structure. In both compounds the a and b lattice parameters were underestimated, indicating that thermal effects may play a role in dictating the distance between the ribbons.

8.3.2 Electronic Properties

Figure 8.3a, c show the band structures of BiSI and BiSeI, respectively, calculated using HSE06+SOC. The inclusion of spin–orbit coupling was tested and found to be essential to accurately describe the electronic structure of both materials (a comparison of the band structures with and without spin–orbit coupling can be found in Fig. C.1 of Appendix C). The relativistic renormalisation of the conduction band

Table 8.1 Lattice parameters of BiSI and BiSeI, with percentage difference from experiment [4, 5] given in parentheses. Lattice parameters and bond lengths are given in Å, with all cell angles found to be 90°. The optimised crystal structures are available online in a public repository [16]

BiSI	a (Å) (%)	b (Å) (%)	c (Å) (%)
PBEsol	8.44 (−0.94)	4.13 (−0.96)	10.26 (+0.79)
PBEsol+D3	8.34 (−2.11)	4.14 (−0.72)	10.05 (−1.28)
PBE	8.81 (+3.40)	4.20 (+0.72)	10.37 (+1.87)
PBE+D3	8.66 (+1.64)	4.21 (+0.96)	11.21 (+10.12)
Experiment	8.52	4.17	10.18
PBEsol	8.63 (−0.80)	4.19 (−0.71)	10.58 (+0.09)
PBEsol+D3	8.53 (−1.95)	4.19 (−0.71)	10.35 (−2.08)
PBE	8.88 (+2.07)	4.28 (+1.42)	10.69 (+1.14)
PBE+D3	8.81 (+1.26)	4.25 (+0.71)	12.03 (+13.81)
Experiment	8.70	4.22	10.57

was found to be over 0.6 eV in both cases. Both band structures show broadly the same features, as expected as BiSI and BiSeI are isoelectronic and isostructural. A primary difference between the band structures is the position of the band edges; in BiSI the valence band maximum (VBM) and conduction band minimum (CBM) appear between Y–Γ and Γ–Z, respectively. In contrast, the VBM of BiSeI is shifted to between Γ–Z, with the CBM situated on Γ. The HSE06+SOC calculated band gaps of BiSI and BiSeI were found to be 1.78 eV and 1.50 eV, respectively. We note that the band gaps of both compounds are strongly dependent on temperature ($dE_g/dT = -7.0 \times 10^{-4}$ eV K^{-1} for BiSI and -6.5×10^{-4} eV K^{-1} for BiSeI), [17] in part due to the weak interactions between ribbons that allow for noticeable thermal expansion. As the HSE06+SOC results fail to account for thermal effects, we have calculated the room temperature (293 K) band gaps using the experimental corrections above. The resulting temperature corrected band gaps of 1.57 eV (BiSI) and 1.31 eV (BiSeI), are in excellent agreement with the room temperature experimental optical band gaps of 1.56–1.59 eV (BiSI) and 1.29–1.32 eV (BiSeI) [17, 18]. As previously observed in the group 15 chalcohalides, [19, 20] the fundamental band gaps are both indirect (E_g^{ind}). However, the direct band gaps (E_g^{dir})—which are more relevant for thin film absorbers due to the need for strong optical absorption—are only slightly larger (∼0.1 eV), indicating that these materials are still ideal for photovoltaic applications.

The ion decomposed density of states (DOS), calculated using HSE06+SOC, is provided in Fig. 8.3b, d. Both show similar features, with the VBM composed of mixing between the I and Ch p states with some Bi s and p states present, and the CBM dominated by Bi p states with the I and Ch p states contributing further into the conduction band. On moving from BiSI to BiSeI, the increase in energy of the Se p states acts to push up the VBM, resulting in a smaller band gap. We note that this shift is in line with the shift in ionisation potential down the chalcogenide group [21].

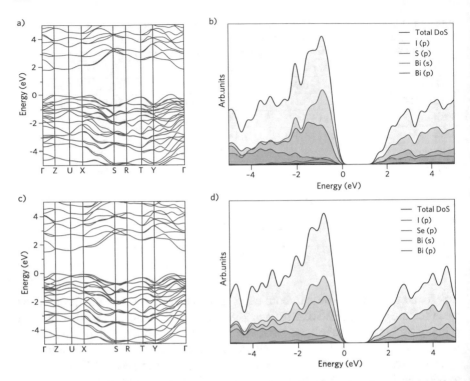

Fig. 8.3 a, c HSE06+SOC calculated band structure and **b, d** partial density of states, for BiSI and BiSeI, respectively. The valence band maximum has been shifted to zero eV in all cases. In panels (**a**) and (**c**), the colours blue and orange denote the valence and conduction bands, respectively

8.3.3   The Bismuth Lone Pair

The appearance of the Bi s and p states toward the top of the valence band can be rationalised through analysis of the Bi lone pair. Similar to many of the heavier post-transition metals, bismuth adopts a valence state two lower than its group valence due to the stability of the $d^{10}s^2p^0$ electronic configuration. Traditionally, lone pair theory predicts that the formation of a stereochemically active lone pair (one that results in a distortion to the crystal structure) is the result of on-site hybridisation of non-bonding Bi s and p states at the VBM [22, 23]. However, in many lone pair systems, such as PbS [24] and CuBiCh$_2$, [25] the metal s and p states are too far apart in energy for hybridisation to occur (~11 eV difference in BiSI). Instead, as detailed in the revised lone pair model by Walsh and Watson [26], the metal s states mix with anion p states, resulting in occupied bonding and anti-bonding states at the top and bottom of the valence band, respectively. This can be seen in the pCOHP analysis provided in Fig. 8.4. An electronic structure of this type, where the VBM

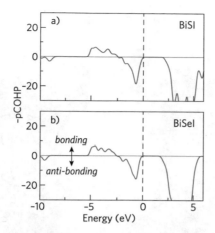

Fig. 8.4 Projected crystal orbital Hamilton population (pCOHP) analysis of **a** BiSI and **b** BiSeI, in which the density of states is partitioned with the sign denoting bonding or anti-bonding character, and the magnitude indicating the strength of the interaction. For each compound the averaged pCOHP across all pairwise interactions in the unit cell is presented. The valence band maximum is set to zero eV

is composed of anti-bonding states, can often result in significantly improved defect tolerance, as dangling bonds formed in the presence of vacancy defects are likely to appear resonant in the valence band [27, 28].

A stereochemically active lone pair will result if the anti-bonding states at the top of the valence band contain a substantial metal s contribution. In this case, these states can mix with the unoccupied metal p orbitals, resulting in a significant stabilisation effect due to the formation of an asymmetric electron density and distortion of the crystal structure [29–31]. As this effect is dependent on the concentration of metal s states at the top of the valence band and this in turn is dictated by the degree of mixing of the anion p and metal s states, the strength of the lone pair is primarily controlled by the relative energies of these states. Metal s and anion p states that are closer in energy will result in stronger lone pairs and, as such, the lone pair effect is seen to lessen when moving down the chalcogen group. When considering heavy metal atoms, such as Bi, which experience strong relativistic effects, the metal $6s$ states are significantly reduced in energy due the scalar contraction of the s-shell, and a further reduction in lone pair strength is observed [26, 32]. Accordingly, the Bi lone pairs are relatively diffuse in BiSI and BiSeI, as can be seen in the charge density isosurface of the VBM shown in Fig. 8.5a. The distortion of the pseudo-octahedra can also be used as a quantitative measure of lone pair strength, and is given by the average deviation in bond angles from the idealised angle of 90° (σ_{oct}). This distortion can be seen to decrease moving from BiSI ($\sigma_{oct} = 7.49°$) to BiSeI ($\sigma_{oct} = 6.61°$), as expected based on the relative strength of the lone pair. In comparison, materials with highly stereochemically active lone pairs, such as Sb_2O_3, possess significantly greater octahedral distortion ($\sigma_{oct} = 21.82°$) [33].

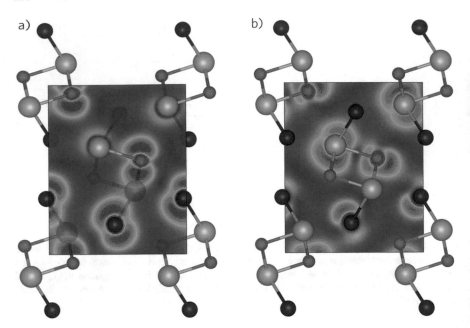

Fig. 8.5 Charge density isosurfaces of **a** the valence band maximum and **b** the conduction band minimum of BiSeI. The colours blue and red indicate areas of low and high electron density, respectively. Bi, Se, and I atoms are represented by grey, green, and purple spheres, respectively

8.3.4 Effective Masses and Dielectric Constants

The effective masses at the band edges of the VBM (m_h^*) and CBM (m_e^*) are provided in Table 8.2. The electron effective masses are significantly larger for the directions between the $[(BiChI)_\infty]_2$ ribbons (between Γ and Z, along [001]), highlighting the weak nature of the dispersive interactions holding the ribbons together. For directions parallel to the ribbons ($\Gamma \rightarrow$ Y, along [100]), the effective masses are much smaller, suggesting that conductivity will be strongly anisotropic and highlighting the potential for significantly increased carrier mobilities in devices containing vertically oriented crystals. BiSeI possesses the smallest electron effective masses, with these occurring in the direction from Γ to Y (0.51 m_0), whereas the smallest electron effective masses in BiSI appear in the direction from the CBM toward Z (0.69 m_0). In both compounds the hole effective masses are small, suggesting that the short hole diffusion lengths seen experimentally in BiSI may be the result of scattering and impurity effects [8]. Interestingly, in both cases the hole effective masses are lower than the electron effective masses. This is similar to in MAPI [9] and likely results from the presence of anti-bonding Bi s states at the VBM. The values obtained for the effective masses are slightly larger than in MAPI ($m_e^* = 0.15\,m_0$; $m_h^* = 0.12\,m_0$) [9] but comparable to other bismuth containing solar absorbers such as Bi_2S_3 ($m_e^* = 0.25\,m_0$; $m_h^* = 0.44\,m_0$) [34].

Table 8.2 Band gaps, effective masses and low-frequency dielectric constants for BiSI and BiSeI, calculated using HSE06+SOC. Effective masses given in units of m_0

	BiSI	BiSeI
E_g^{ind} (eV)	1.78	1.50
E_g^{dir} (eV)	1.82	1.60
m_e^* (CBM → Z)	0.69	4.52
m_e^* (CBM → Γ)	1.78	–
m_e^* (CBM → Y)	–	0.51
m_h^* (VBM → Γ)	0.51	–
m_h^* (VBM → Y)	0.36	–
m_h^* (VBM → Z)	–	0.40
m_h^* (VBM → Γ)	–	0.28
ε_r	36.8	35.8

As previously mentioned, large dielectric constants (ε_r) are considered a primary factor in the excellent performance seen in the hybrid perovskites [3, 9, 35]. The dielectric response of a material is crucial in determining the charge capture cross-section of defects, with more effective charge screening resulting in smaller cross sections and lower recombination and scattering rates [36]. Additionally, larger dielectric constants should promote smaller defect binding energies (in effective mass theory $E_b = m_e^*/2\varepsilon_r^2$), [9, 37] preventing defect states from appearing deep within the gap. As such, we have calculated the dielectric constants of both materials, finding them to be 36.8 and 35.8 for BiSI and BiSeI, respectively (Table 8.2). While smaller than those seen in the lead hybrid halide perovskites—where dielectric constants can reach 60–70 or higher, thanks to screening provided by the organic cation [10, 38]—they are noticeably larger than in other established absorbers such as CZTS ($\sim$9) [39]. We therefore expect this combination of small effective masses and large dielectric constants to promote shallower defect states and reduced levels of charge-carrier recombination [40].

8.3.5 Alignments with Electron and Hole Contact Layers

Band alignments calculated using HSE06+SOC are provided in Fig. 8.6. The ionisation potentials (IPs) of BiSI and BiSeI are significantly larger than many established photovoltaic absorbers (6.4 eV and 6.2 eV below the vacuum level, respectively) [9, 41, 42]. It is clear that the band gap reduction when moving from BiSI to BiSeI results from the smaller ionisation potential of BiSeI, combined with concomitant stabilisation of the conduction band minimum (an increase in the electron affinity). Figure 8.6 also details the band alignments obtained experimentally by Hahn et al., which show an increase in the ionisation potential when S is replaced with Se (from

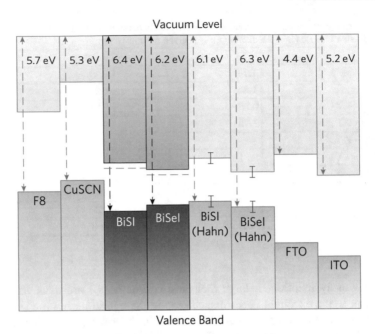

Fig. 8.6 HSE06+SOC calculated band alignment of BiSI and BiSeI with a series of transparent conducting oxides and hole transporting materials, taken from experiment [46–51]. Electron affinities calculated using the experimental band gap and HSE06+SOC ionisation potentials are show in dashed red lines. Experimental band alignments performed by Hahn et al. are also plotted for comparison [7]. Experimental uncertainty is indicated by error bars

∼6.1 eV in BiSI to ∼6.3 in BiSeI) [7]. This trend is unexpected as the increase in energy of the Ch p orbital down the chalcogen group typically results in a raising of the valence band maximum and decrease in the ionisation potential, as observed in many other mixed chalcogenide systems including $CdS_{1-x}Se_x$, [43] $CdSe_xTe_{1-x}$, [44] and $Cu_2ZnSn(S,Se)_4$ [45]. This discrepancy may be rationalised through the authors' observation of BiOI contamination on the surface of their samples [7].

Band alignments for a range of commonly used transparent conducting oxides (TCOs) and hole transporting materials (HTMs) are also included in Fig. 8.6. It is clear that the materials used in the photovoltaic devices constructed by Hahn et al. (namely FTO and p-CuSCN) result in band misalignments against both the TCO and HTM layers [7]. For example, when the electron affinity of BiSI is calculated based on the HSE06+SOC ionisation potential and experimentally measured band gap, the maximum obtainable open-circuit voltage from Hahn et al.'s device is limited to just ∼0.4 eV [46, 47]. An alternative HTM with a lower ionisation potential, such as poly(fluorene-2,7-diyl) (F8, IP = 5.7 eV), [48] will therefore allow for significantly larger open-circuit voltages and increased power conversion efficiencies. When considering the TCO layer, the electron affinities of BiSI and BiSeI (EA = 4.9 eV and 5.0 eV, respectively) are larger than that of FTO (EA ≃ 4.4 eV), [49] indicating that electron transfer from the absorber to the n-type layer will be poor. TCOs with larger electron affinities—for example indium tin oxide (ITO, EA $\simeq 5.2\,\text{eV}$) [50, 51]— should therefore be employed to enable more efficient transport of photo-generated carriers. We note that the above analysis relies solely on the fundamental band alignments of the bulk crystal and does not take into account effects originating from interfacial strain and other chemical interactions. Regardless, it is clear that the use of revised contact materials will be essential if device efficiencies are to improve.

8.3.6 Defect Chemistry

As BiSI and BiSeI show promising electronic properties for photovoltaic applications, we have performed calculations to investigate how their defect behaviour will influence device performance. The chemical potential phase diagram of both compounds is provided in Fig. 8.7. The thermodynamically accessible range of chemical potentials are bounded by the formation of the binary bismuth chalcogenides (Bi_2S_3 and Bi_2Se_3) and BiI_3, resulting in a narrow but broad region spanning a significant portion of chemical potential space. In each case, two synthetic environments were identified: the most p-type environment, A, shows bismuth-poor and anion-rich (sulfur and iodine-rich) conditions; whereas the most n-type, B, is anion-poor and bismuth-rich.

In this study, the n-type defects investigated were anion vacancies (V_I and V_{Ch}), iodine on chalcogen antisite (I_{Ch}), anion on bismuth antisites (Ch_{Bi}, and I_{Bi}), and bismuth interstitials (Bi_i). For p-type defects, we have considered a bismuth vacancy (V_{Bi}), chalcogen on iodine antisites (Ch_I), bismuth on anion antisites (Bi_{Ch} and Bi_I), and anion interstitials (Ch_i and I_i). Two interstitial defect sites were located in the voids between the one-dimensional ribbons, comprising one penta-coordinated site and one octa-coordinated site (Fig. 8.8).

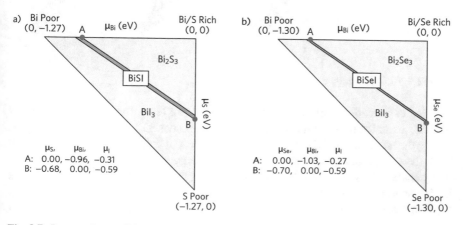

Fig. 8.7 Range of accessible chemical potentials for **a** BiSI and **b** BiSeI

Fig. 8.8 Crystal structure of the $2 \times 3 \times 1$ supercell of BiSI used in defect calculations, indicating the **a** penta-coordinated and **b** octa-coordinated interstitial sites. Bi, S, and I atoms are denoted by grey, green, and purple spheres, respectively. Interstitial atom denoted by a blue sphere

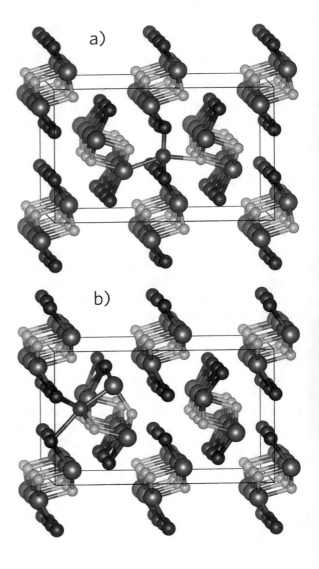

8.3.6.1   BiSI Defects

The defect formation energies as a function of Fermi level, for the intrinsic defects of BiSI, under the most p-type conditions (environment A), are provided in Fig. 8.9. Due to the large number of defect states, the interstitial defects are plotted separately from the vacancies and antisites. The lowest energy acceptor, S_I, possesses a relatively shallow $0/-1$ transition level 0.14 eV above the valence band maximum. This is compensated by the lowest energy donor defects, I_S and V_I, which will act to trap the Fermi level in the middle of the band gap. This is confirmed by calculation of

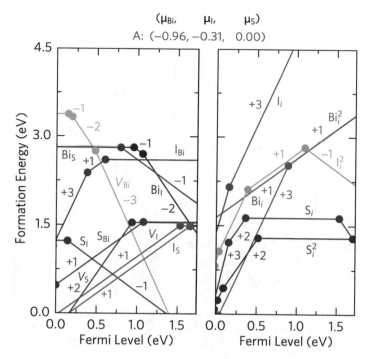

Fig. 8.9 Defect formation energies as a function of Fermi level for BiSI under a p-type (Bi-poor, S and I-rich) chemical potential environment. The slope of the lines denotes the charge state, with a steeper line indicating a higher charge state. The solid dots represent the transition levels, $\varepsilon(q/q')$

the self-consistent Fermi level (at a temperature of 21 °C), which indicates the Fermi level will be pinned 0.77 eV above the valence band maximum, resulting in a hole concentration of 1.21×10^8 cm^{-3}.

The remainder of the acceptor antisite defects possess large formation energies (around 3 eV for the neutral charge states), which will preclude their formation at the low processing temperatures required to synthesise BiSI thin films [52]. Considering the interstitial defects: I_i^2, S_i, and S_i^2 may act as acceptors provided the Fermi level sits close to the conduction band minimum. As the Fermi level under environment A will be trapped in the middle of the band gap, these defects will therefore not contribute to p-type conductivity.

The S_{Bi} and S_i^2 donor defects show ultra-deep states in the middle of the gap, which could contribute to trap-assisted charge-carrier recombination. Analysis of the self-consistent Fermi level, however, reveals that these defects will be present at remarkably low concentrations (5.8×10^1 cm^{-3} and 0.8×10^1 cm^{-3}, respectively, calculated at a representative device operating temperature of 21 °C) and will therefore have a limited effect on p-type conductivity.

The transition level diagram for the intrinsic defects of BiSI under the most n-type conditions (environment B), is provided in Fig. 8.10. I_S is the lowest energy donor

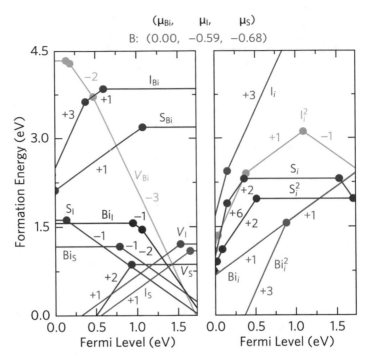

Fig. 8.10 Defect formation energies as a function of Fermi level for BiSI under an *n*-type (Bi-rich, S and I-poor) chemical potential environment. The slope of the lines denotes the charge state, with a steeper line indicating a higher charge state. The solid dots represent the transition levels, $\varepsilon(q/q')$

defect, with a shallow $0/+1$ transition level 0.08 eV beneath the conduction band minimum. The lowest energy acceptor defect, S_I^{-1}, again causes the Fermi level to be trapped mid-gap (0.57 eV below the conduction band edge). V_I is also a low energy donor but possesses a deeper $0/+1$ transition state 0.20 eV from the conduction band minimum. The neutral V_I is higher in energy than the neutral V_S, however, the sulfur vacancy is a negative-U defect, possessing an ultra-deep $0/+2$ transition state in the middle of the band gap. Examination of the charge-density isosurfaces for each charge state reveals the origin of this behaviour: in the neutral state, the two excess electrons are delocalised over the three bismuth atoms closest to the defect site, with the bulk of the electron density occupying the void in the structure (Fig. 8.11a). This arrangement is stabilised by a slight displacement of the bismuth atom opposite the defect site, which is pulled forward into the vacancy, allowing the rest of the one-dimensional ribbon to remain largely unperturbed. Moving to the $+1$ charge state, the remaining electron is localised solely on the bismuth atom opposite the defect site, with significantly less electron density occupying the void (Fig. 8.11b). As the two bismuth atoms either side of the vacancy are no longer hybridised, they are pulled outward, causing a high-energy structural distortion that travels the length of the one-dimensional chain, thereby increasing the strain in the lattice. In the $+2$

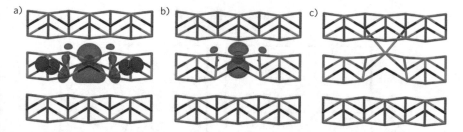

Fig. 8.11 Crystal structure and charge density isosurfaces of the V_S defect in the **a** neutral, **b** +1, and **c** +2 charge states, viewed along the [001] direction. Green, grey and purple cylinders indicate bonding to sulfur, bismuth and iodine atoms, respectively. Electron density is represented in orange. The isosurface level was set to $0.03\,\mathrm{eV}\,\text{Å}^{-3}$

charge state, the bismuth opposite the vacancy site is displaced backwards as it bonds with two iodine atoms in an nearby ribbon (Fig. 8.11c). Overall, the distortion of this geometry is comparable to the +1 charge state, however, the additional bonding provides greater structural stability.

Bi_i^2 is the lowest energy donor interstitial, possessing an ultra-deep $+1/+3$ transition state, however, its high formation energy at the self-consistent Fermi level indicates it will only be present in very low concentrations. It should be noted that both V_S and Bi_S are low energy, ultra-deep defects that may contribute to charge-carrier recombination. As such, post-annealing in a sulfur atmosphere should be considered to reduce the effect on n-type conductivity.

8.3.6.2 BiSeI Defects

The defect chemistry of BiSeI is similar to that of its sulfur counterpart. The defect formation energies as a function of Fermi level, for the intrinsic defects of BiSeI, under the most p-type conditions (environment A), is provided in Fig. 8.12. Se_I is the lowest energy acceptor defect, possessing a transition level 0.13 eV above the valence band maximum. In comparison to the equivalent defect in BiSI (S_I), Se_I is both lower in energy and more shallow. The lowest energy donor defect is Se_I, which will trap the Fermi level 0.68 eV above the valence band, resulting in a hole concentration of $8.52 \times 10^7\,\mathrm{cm}^{-3}$ at 21 °C.

The main variation between the BiSeI and BiSI p-type defects is found for Se_{Bi}, which possesses a significantly reduced formation energy (enabling high defect concentrations of $4.77 \times 10^{10}\,\mathrm{cm}^{-3}$) and an ultra-deep $0/+1$ transition level. As such, this defect may play a role in trap-assisted recombination. Similar to BiSI, the remaining interstitial and antisite defects show significantly higher formation energies, which preclude their formation in high concentrations.

The transition level diagram for the intrinsic defects of BiSeI under the most n-type conditions (environment B) is provided in Fig. 8.13. In general, the larger electron affinity of BiSeI results in donor defects that are more shallow than in

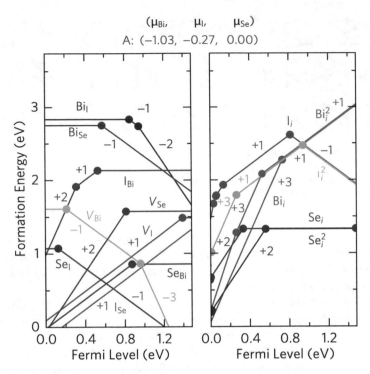

Fig. 8.12 Defect formation energies as a function of Fermi level for BiSeI under a p-type (Bi-poor, Se and I-rich) chemical potential environment. The slope of the lines denotes the charge state, with a steeper line indicating a higher charge state. The solid dots represent the transition levels, $\varepsilon(q/q')$

the sulfur analogue. I_{Se} is the lowest energy donor defect, possessing a transition level resonant in the conduction band (0.17 eV above the conduction band edge). This is compensated by both Bi_{Se} and Se_I, causing the Fermi level to be pinned 0.43 eV beneath the conduction band minimum, producing an electron concentration of 7.64×10^{11} cm^{-3} at 21 °C.

Similar to BiSI, both Bi_{Se} and V_{Se} possess ultra-deep transition levels, with relatively low formation energies and self-consistent concentrations of 4.84×10^{12} cm^{-3} and 1.17×10^{7} cm^{-3}, respectively. As such, annealing in a selenium atmosphere may be required to alleviate any negative effects on charge-carrier recombination. In all cases, the interstitial defects possess large formation energies, and will not be present in high concentrations at standard device operating conditions.

8.3.6.3   Defect Engineering

Under both n- and p-type conditions, the Fermi level of BiSI and BiSeI will be trapped mid gap. As such, both compounds will behave as *intrinsic* semiconductors and are most well suited to use in a p–i–n junction solar cell architecture. The transition level

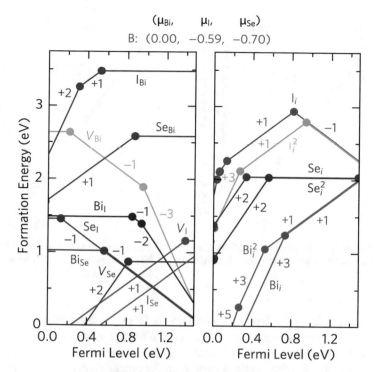

$(\mu_{Bi}, \quad \mu_{I}, \quad \mu_{Se})$
B: (0.00, −0.59, −0.70)

Fig. 8.13 Defect formation energies as a function of Fermi level for BiSeI under an *n*-type (Bi-rich, Se and I-poor) chemical potential environment. The slope of the lines denotes the charge state, with a steeper line indicating a higher charge state. The solid dots represent the transition levels, $\varepsilon(q/q')$

diagrams reveal the existence of low energy deep defects (for example Ch_{Bi} and V_{Ch}), present in reasonably high concentrations, at the limits of both *n*- and *p*-type synthesis conditions. As previously discussed, these defect sites are likely to contribute to trap-assisted recombination, and may have a detrimental effect on the open circuit voltage (see Sect. 1.2.2). In order to minimise the effect of these defects on recombination, we have explored the defect behaviour across the full range of achievable chemical potentials. Using the self-consistent Fermi level method, we have calculated the charge-carrier and defect concentrations for each point within the region of stability, identified in the chemical potential phase diagrams shown in Fig. 8.7. Due to the low carrier concentrations present, Shockley–Read–Hall recombination will be the dominant recombination mechanism in BiSI and BiSeI devices [53]. Accordingly, we have calculated an estimate of the recombination rate, at each set of chemical potentials, using the Shockley–Read–Hall recombination model discussed in Sect. 3.4.6. Here, the recombination rate, R^{SRH}, is dependent on the concentration of holes and electrons, in addition to the number of defects and the positions of the charge transition levels.

In this model, defect states closer to the centre of the band gap show exponentially greater activity as recombination centres than states located close to the valence

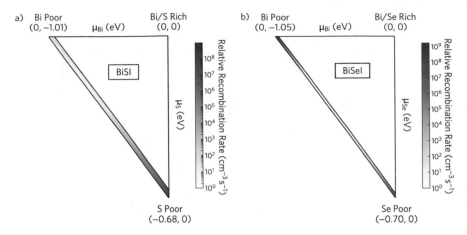

Fig. 8.14 Shockley–Read–Hall recombination rate (R_{SRH}) across stability region in chemical potential space for **a** BiSI and **b** BiSeI. Larger recombination rates are indicated by darker red regions

band maximum or conduction band minimum. In this way, as the number of shallow *versus* deep defects changes across chemical potential space, the overall rate of recombination will also vary. As calculating defect charge-capture cross sections from first-principles is extremely computationally demanding, [54] we have instead assumed a constant cross-section for all defects ($\sigma = 1 \times 10^{-15}$ cm^2). Accordingly, our results will form an qualitative estimate of the true recombination rate.

The Shockley–Read–Hall recombination rate, as a function of chemical potential, is provided in Fig. 8.14. We have normalised the data to the region of lowest recombination. Considering BiSI: the p-type conditions possess the lowest levels of recombination, resulting from the high defect formation energy of the most prevalent deep defects (V_S and S_{Bi}). The recombination rate increases gradually as the chemical potentials shift to favouring the formation of n-type defects, due to the reduction in formation energy of the ultra-deep Bi$_S$ and V_S defects. Under the limits of n-type conductivity, the recombination rate is over 10^7 s^{-1} more than under p-type conditions. As such, bismuth poor synthesis conditions should be considered to minimise the levels of recombination, and maintain an optimal open-circuit voltage.

In contrast, BiSeI shows high levels of recombination under the limits of both n- and p-type synthesis conditions. This can be attributed to the low formation energy of the ultra-deep V_{Se} and Bi$_{Se}$ defects—which dominate under the most n-type environment—and the Se$_{Bi}$ defect—which is present under the most p-type environment. In the middle of the chemical potential space, however, the recombination rate is reduced considerably, due to the absence of any low-energy deep defect states. Accordingly, use of a more stoichiometric ratio of starting materials will likely limit the formation of "killer-defects" in high concentrations.

8.4 Conclusions

In this chapter, we have investigated BiSI and BiSeI as potential solar absorber materials. Using first-principles relativistic electronic structure theory, we show that BiSI and BiSeI possess ideal electronic structures for photovoltaic devices, with band gaps in the optimal range specified by the Shockley–Queisser limit. Small effective masses in the directions parallel to the one-dimension ribbons, coupled with large dielectric constants, are expected to promote charge transport and reduce the effects of charged-defect scattering. Based on analysis of the electronic band alignments, we have identified band misalignments as a potential cause of the poor efficiencies of BiSI and BiSeI devices previously reported in the literature. Accordingly, we have suggested alternative electron and hole contact materials that we expect will result in improved open-circuit voltages, charge-carrier extraction, and device efficiencies. By calculating the defect properties of both compounds, we identify these materials as intrinsic semiconductors best suited for use in a p–i–n junction solar cell architecture. Through mapping the rate of Shockley–Read–Hall recombination across chemical potential space, we further reveal regions showing reduced recombination rates. As such, judicious choice of synthesis conditions will be essential in producing devices with optimal trap-assisted recombination properties. These results will be beneficial for experimental groups aiming to produce efficient bismuth chalcohalide photovoltaic devices in the future.

Notes

The work discussed in this chapter formed the basis for two publications:

1. Ganose AM, Butler KT, Walsh A, Scanlon DO (2016) Relativistic electronic structure and band alignment of BiSI and BiSeI: candidate photovoltaic materials. J Mater Chem A 4:2060 (2016)
2. Ganose AM, Matsumoto S, Buckeridge J, Scanlon DO (2018) Defect engineering of earth-abundant solar absorbers BiSI and BiSeI. Chem Mater 30:3827–3835

The optimised crystal structures, for all compounds discussed in this chapter, are provided in an online repository [16].

References

1. Ganose AM, Butler KT, Walsh A, Scanlon DO (2016) Relativistic electronic structure and band alignment of BiSI and BiSeI: candidate photovoltaic materials. J Mater Chem A 4:2060–2068
2. Ganose AM, Matsumoto S, Buckeridge J, Scanlon DO (2018) Defect engineering of earth-abundant solar absorbers BiSI and BiSeI. Chem Mater 30:3827–3835

3. Brandt RE, Stevanović V, Ginley DS, Buonassisi T (2015) Identifying defect-tolerant semi-conductors with high minority-carrier lifetimes: beyond hybrid lead halide perovskites. MRS Commun 5:1–11
4. Demartin F, Gramaccioli C, Campostrini I (2010) Demicheleite-(I), BiSI, a New Mineral from La Fossa Crater, Vulcano, Aeolian Islands. Italy Mineral Mag 74:141–145
5. Braun TP, DiSalvo FJ (2000) Bismuth selenide iodide. Acta Crystallogr Sect C: Cryst Struct Commun 56:1–2
6. Zhou Y, Wang L, Chen S, Qin S, Liu X, Chen J, Xue D-J, Luo M, Cao Y, Cheng Y (2015) Thin-Film Sb$_2$Se$_3$ photovoltaics with oriented one-dimensional ribbons and benign grain boundaries. Nat Photon 9:409–415
7. Hahn NT, Rettie AJ, Beal SK, Fullon RR, Mullins CB (2012) n-BiSI thin films: selenium doping and solar cell behavior. J Mater Chem C 116:24878–24886
8. Hahn NT, Self JL, Mullins CB (2012) BiSI micro-rod thin films: efficient solar absorber electrodes? J Phys Chem Lett 3:1571–1576
9. Frost JM, Butler KT, Brivio F, Hendon CH, Van Schilfgaarde M, Walsh A (2014) Atomistic origins of high-performance in hybrid halide perovskite solar cells. Nano Lett 14:2584–2590
10. Walsh A (2015) Principles of chemical bonding and band gap engineering in hybrid organic-inorganic halide perovskites. J Mater Chem C 119:5755–5760
11. Kunioku H, Higashi M, Abe R (2016) Low-temperature synthesis of Bismuth Chalcohalides: candidate photovoltaic materials with easily. Continuously Controllable Band gap Sci Rep 6:32664
12. Pulay P (1969) Ab initio calculation of force constants and equilibrium geometries in poly-atomic molecules: I. Theory Mol Phys 17:197–204
13. Dronskowski R, Blöchl PE (1993) Crystal Orbital Hamilton Populations (COHP): energy-resolved visualization of chemical bonding in solids based on density-functional calculations. J Phys Chem 97:8617–8624
14. Maintz S, Deringer VL, Tchougréeff AL, Dronskowski R (2013) Analytic projection from plane-wave and paw wavefunctions and application to chemical-bonding analysis in Solids. J Comput Chem 34:2557–2567
15. Wei S-H, Zunger A (1998) Calculated natural band offsets of all II-VI and III-V semiconductors: chemical trends and the role of cation d orbitals. Appl Phys Lett 72:2011–2013
16. https://github.com/SMTG-UCL/BiChI. Accessed 2015 Nov 2012
17. Chepur D, Bercha D, Turyanitsa I, Slivka VY (1968) Peculiarities of the energy spectrum and edge absorption in the chain compounds A^{V}B^{VI}C^{VII}. Phys Status Solidi B 30:461–468
18. Shin D-W, Hyun S-C, Park S-A, Kim Y-G, Kim C-D, Kim W-T (1994) Optical properties of undoped and Ni-doped V^A-VI^A-VII^A single crystals. J Phys.Chem Solids 55:825–830
19. Park S-A, Kim M-Y, Lim J-Y, Park B-S, Koh J-D, Kim W-T (1995) Optical properties of undoped and V-Doped V^A-VI^A-VII^A single crystals. Phys Status Solidi B 187:253–260
20. Hyun S-C, Kim Y-G, Kim M-Y, Koh J-D, Park B-S, Kim W-T (1995) Optical properties of undoped and Chromium-DopedV^A-VI^A-VII^A single crystals. J Mater Sci 30:6113–6117
21. Butler KT, Frost JM, Walsh A (2015) Band alignment of the hybrid halide perovskites CH$_3$NH$_3$PbCl$_3$, CH$_3$NH$_3$PbBr$_3$ and CH$_3$NH$_3$PbI$_3$. Mater Horiz 2:228–231
22. Dunitz J, Orgel L (1960) Stereochemistry of ionic solids. Adv Inorg Chem Radiochem 2:1–60
23. Orgel L (1959) The stereochemistry of b subgroup metals. Part II. The Inert Pair. J Chem. Soc 769:3815–3819
24. Walsh A, Watson GW (2005) The origin of the stereochemically active Pb (II) lone pair: DFT calculations on PbO and PbS. J Solid State Chem 178:1422–1428
25. Temple DJ, Kehoe AB, Allen JP, Watson GW, Scanlon DO (2012) Geometry, electronic structure, and bonding in CuMCh$_2$ (M= Sb, Bi; Ch= S, Se): alternative solar cell absorber materials? J Mater Chem C 116:7334–7340
26. Walsh A, Payne DJ, Egdell RG, Watson GW (2011) Stereochemistry of post-transition metal oxides: revision of the classical lone pair model. Chem Soc Rev 40:4455–4463
27. Zhang S, Wei S-H, Zunger A, Katayama-Yoshida H (1998) Defect Physics of the CuInSe$_2$ chalcopyrite semiconductor. Phys Rev B 57:9642

28. Zakutayev A, Caskey CM, Fioretti AN, Ginley DS, Vidal J, Stevanovic V, Tea E, Lany S (2014) Defect tolerant semiconductors for solar energy conversion. J Phys Chem Lett 5:1117–1125
29. Walsh A, Watson GW (2005) Influence of the anion on lone pair formation in Sn (II) monochalcogenides: a DFT study. J Phys Chem B 109:18868–18875
30. Walsh A, Watson GW, Payne DJ, Edgell RG, Guo J, Glans P-A, Learmonth T, Smith KE (2006) Electronic structure of the α and δ phases of Bi_2O_3: a combined Ab Initio and X-ray spectroscopy Study. Phys Rev B 73:235104
31. Waghmare U, Spaldin N, Kandpal H, Seshadri R (2003) First-principles indicators of metallicity and cation off-centricity in the IV–VI Rocksalt chalcogenides of divalent Ge, Sn, and Pb. Phys Rev B 67:125111
32. Walsh A, Watson GW (2007) Polymorphism in Bismuth Stannate: a first-principles study. Chem Mater 19:5158–5164
33. Svensson C (1974) The crystal structure of orthorhombic antimony trioxide, Sb_2O_3. Acta Crystallogr Sect B: Struct Sci 30:458–461
34. Guo D, Hu C, Zhang C (2013) First-principles study on doping and temperature dependence of thermoelectric property of Bi_2S_3 thermoelectric material. Mater Res Bull 48:1984–1988
35. Walsh A, Scanlon DO, Chen S, Gong X, Wei S-H (2015) Self-regulation mechanism for charged point defects in hybrid halide perovskites. Angew Chem 127:1811–1814
36. Bube RH (1992) Photoelectronic properties of semiconductors. Cambridge University Press
37. Cardona M, Peter YY (2005) Fundamentals of semiconductors. Springer
38. Frost JM, Butler KT, Walsh A (2014) Molecular ferroelectric contributions to anomalous hysteresis in hybrid perovskite solar cells. APL Mater 2:081506
39. Fernandes P, Salomé P, Da Cunha A (2009) Precursors' order effect on the properties of sulfurized Cu_2ZnSnS_4 thin films. Semicond Sci Technol 24:105013
40. Allan G, Delerue C, Lannoo M, Martin E (1995) Hydrogenic impurity levels, dielectric constant, and coulomb charging effects in silicon crystallites. Phys Rev B 52:11982
41. Burton LA, Walsh A (2013) Band alignment in SnS thin-film solar cells: possible origin of the low conversion efficiency. Appl Phys Lett 102:132111
42. Jaegermann W, Klein A, Mayer T (2009) Interface engineering of inorganic thin-film solar cells-materials-science challenges for advanced physical concepts. Adv Mater 21:4196–4206
43. Chen Z, Peng W, Zhang K, Zhang J, Yang X, Numata Y, Han L (2014) Band alignment by ternary crystalline potential-tuning interlayer for efficient electron injection in quantum dot-sensitized solar cells. J Mater Chem A 2:7004–7014
44. MacDonald BI, Martucci A, Rubanov S, Watkins SE, Mulvaney P, Jasieniak JJ (2012) Layer-by-layer assembly of sintered $CdSe_xTe_{1-x}$ nanocrystal solar cells. ACS Nano 6:5995–6004
45. Chen S, Walsh A, Yang J-H, Gong X, Sun L, Yang P-X, Chu J-H, Wei S-H (2011) Compositional dependence of structural and electronic properties Of $Cu_2ZnSn(S, Se)_4$ alloys for thin film solar cells. Phys Rev B 83:125201
46. Boix PP, Larramona G, Jacob A, Delatouche B, Mora-Seró I, Bisquert J (2011) Hole transport and recombination in all-solid Sb_2S_3-sensitized TiO_2 solar cells using CuSCN As hole transporter. J Mater Chem C 116:1579–1587
47. Chao X, Lei C, Hongchun Y (2013) Study on the synthesis, characterization of p-CuSCN/n-Si heterojunction. Open Mater Sci J 7:29–32
48. Inbasekaran M, Woo E, Wu W, Bernius M, Wujkowski L (2000) Fluorene homopolymers and copolymers. Synth Met. 111:397–401
49. Helander M, Greiner M, Wang Z, Tang W, Lu Z (2011) Work function of fluorine doped tin oxide. J Vac Sci Technol A 29:011019
50. Dong C, Yu W, Xu M, Cao J, Chen C, Yu W, Wang Y (2011) Valence band offset of Cu_2O/In_2O_3 heterojunction determined by X-ray photoelectron spectroscopy. J Appl Phys 110:073712
51. Höffling B, Schleife A, Rödl C, Bechstedt F (2012) Band discontinuities at Si–TCO interfaces from quasiparticle calculations: comparison of two alignment approaches. Phys Rev B 85:035305

52. Brandt RE, Poindexter JR, Gorai P, Kurchin RC, Hoye RL, Nienhaus L, Wilson MW, Polizzotti JA, Sereika R, Žaltauskas R, Lee LC, MacManus-Driscoll JL, Bawendi Stevanović V, Buonassisi T (2017) Searching for "defect-tolerant" photovoltaic materials: combined theoretical and experimental screening. Chem Mater 29:4667–4674
53. Shockley W, Read W Jr (1952) Statistics of the recombinations of holes and electrons. Phys Rev 87:835
54. Alkauskas A, Dreyer CE, Lyons JL, Van de Walle CG (2016) Role of excited states in Shockley-Read-Hall recombination in wide-band-gap semiconductors. Phys Rev B 93:201304

Summary

The aim of this thesis was to critically assess the fundamental properties of emerging photovoltaic absorbers through use of first-principles electronic structure calculations. It is currently impossible to work in the field of photovoltaics without acknowledging the success of the hybrid perovskites, which have seen truly impressive efficiency gains in the last nine years. Despite considerable research attention, however, the hybrid perovskites present several challenges which remain to be overcome. The materials examined in this thesis have all emerged in response to these challenges. In order to evaluate the potential of these materials as photovoltaic absorbers, an in-depth understanding of their atomic-scale properties is essential. In practice, this means not only understanding bulk optoelectronic properties but also the role of defects, which will play a major role on carrier transport and recombination. In this chapter, we summarise the major findings from our work and highlight future research directions.

In Part 2 of this thesis we looked at absorbers derived from modifications to the perovskite structure. In recent months, the layered compound, $(CH_3NH_3)_2Pb(SCN)_2I_2$ (MAPSI), has been proposed as a more stable alternative to the cubic perovskite $CH_3NH_3PbI_3$, however the origin of this stability was unknown. By examining the energetics of decomposition for both compounds, we demonstrated that MAPSI is intrinsically more stable against decomposition via several routes. To investigate MAPSI's photovoltaic performance, we performed relativistic density functional theory calculations. These revealed MAPSI possesses a band gap suitable for a top cell in a photovoltaic tandem device, along with small hole and electron effective masses that should enable high carrier mobilities. Furthermore, the majority of intrinsic vacancy defects are shallow and therefore should not adversely affect carrier recombination rates.

We demonstrated that, similar to the ABX_3 perovskites which can be tuned on the A, B and X sites, MAPSI can also act as a parent compound to a range of MAPSI-structured analogues. Replacing SCN with OCN and SeCN, I with Cl and Br, and Pb with Sn, enables a series of compounds with a diverse range of electronic

A. Ganose, *Atomic-Scale Insights into Emergent Photovoltaic Absorbers*, Springer Theses, https://doi.org/10.1007/978-3-030-55708-9

properties. The band gap, valence bandwidth, and charge-carrier effective masses of these analogues are determined by the relative energies of the halide and pseudohalide p states. Furthermore, all analogues are stable with respect to decomposition. The iodide analogues all possess band gaps suitable for photovoltaic top cells and are therefore of interest to the photovoltaic community. In particular, the tin iodides present an exciting possibility for the future of stable lead-free perovskites. We therefore welcome the experimental realisation of these promising materials. Future work may focus on the possibility of quasi-2D MAPSI structured perovskites. Indeed, as layered perovskites show dramatic efficiency and stability improvements when the layer thickness is greater than 1 perovskite sheet, so too may the MAPSI analogues. This may provide a mechanism to further lower the band gap and bring it into the range suitable for single-junction photovoltaics.

The toxicity of lead remains the second major concern surrounding the hybrid perovskites. In Chap. 6 we investigated the lead-free vacancy-ordered defect perovskites Cs_2SnI_6 and Cs_2TeI_6. We demonstrated that while Cs_2SnI_6 possesses a direct band gap, it is dipole disallowed and therefore results in weak optical absorption. Accordingly, we provided an explanation for the apparent mismatch between theory and experiment previously discussed in the literature. Furthermore, the fundamental band gap of Cs_2TeI_6 is indirect and so neither compounds are expected to perform well as photovoltaic absorbers. Despite this, we showed that both contain intriguing electronic structures deserving of further study. In particular, the presence of dispersive conduction band states containing significant Sn contributions, despite the presence of regular vacancies, is due to inter-octahedral I–I interactions that span the voids in the lattice.

We showed that the dramatic reduction in conductivity seen by our experimental collaborators upon tellurium substitution in the series $Cs_2Sn_{1-x}Te_xI_6$, cannot be explained by differences in carrier effective masses. Instead, this behaviour is due to the depth of iodine vacancy defects, which are shallowest for Cs_2SnI_6 and become deeper with increasing tellurium incorporation. To overcome the primary limitation of these materials in the context of photovoltaics—i.e. the dipole disallowed fundamental band gap—it may be fruitful to investigate $Cs_2SnI_xBr_{6-x}$ solid-solutions, with the aim of breaking the symmetry of the frontier orbitals and restoring strong optical absorption.

In Part 3 we looked for photovoltaic absorbers beyond the hybrid perovskite family. Bismuth-based materials were identified as attractive candidate absorbers due to their earth-abundant and non-toxic nature. We demonstrated that BiSI and BiSeI possess ideal electronic structures for photovoltaic devices, with band gaps in the optimal range specified by the Shockley–Queisser limit. Their small effective masses and large dielectric constants should further promote efficient charge transport and reduce the effects of charged-defect scattering. We demonstrated that band misalignments are most likely the cause of the poor efficiencies of BiSI and BiSeI devices previously reported in the literature. Furthermore, we calculated their defect properties and showed both compounds are best suited for use in p–i–n junction solar cell architectures. Lastly, we demonstrated that judicious choice of synthesis conditions should enable dramatic reductions in the rate of Shockley–Read–Hall recombination.

We believe these results will aid the fabrication of more efficient bismuth chalco-halide photovoltaic devices. Future theoretical work could focus on the interface between the absorber and device contact layers, with the aim of minimising the effects of surface recombination. Additionally, it may be beneficial to investigate the solid solution series Bi(S,Se)I to enable tuning of the electronic, optical, and defect properties for photovoltaic applications.

In the process of performing this work, the field of emerging photovoltaics has seen some dramatic changes. The efficiency of perovskite photovoltaics has risen from ~15% to over 23%. Simultaneously, the development of the layered perovskites as stable alternatives has attracted considerable attention, with most of the best performing perovskite devices now containing a small fraction of the layered analogues. Despite this, the stability of the layered perovskites is not properly understood and future work investigating their tolerance to water is clearly needed. The results presented in this thesis provide an insight into this class of layered materials. Additional work on MAPSI and the Ruddlesden–Popper phases will be necessary to optimise the performance of devices containing these materials. In contrast, while the vacancy-ordered double perovskites initially showed promise due to their ideal optical band gaps, the work in this thesis indicates they posses small fundamental band gaps and are undesirable as solar absorbers. Regardless, they still present interesting toy systems to better understand the ability of iodine anions to provide dispersive conduction band states.

The last few years have also seen the rise of lone-pair materials as promising photovoltaics. These absorbers, containing Bi^{3+} and Sb^{3+} cations, are expected to show many similar properties to the hybrid perovskites. Our investigation into one such material, BiSI, reveals it posses optimal optoelectronic properties, but complicated defect behaviour that must be considered during device fabrication. This is in contrast to MAPI, which shows high performance despite relatively high defect concentrations. As such, further work will be necessary to determine whether the defect tuning proposed in this thesis will translate into more efficient devices.

Appendix A
Pseudohalide Perovskite Absorbers

See Figs. A.1, A.2, A.3 A.4 and Table A.1.

Fig. A.1 Effect of spin–orbit coupling on the band structure of $(CH_3NH_3)_2Pb(SCN)_2I_2$. The HSE+SOC band structure is shown by black lines, with the HSE only band structure shown via dashed red lines. The valence band maxmimum of the HSE+SOC band structure is set to $0\,eV$

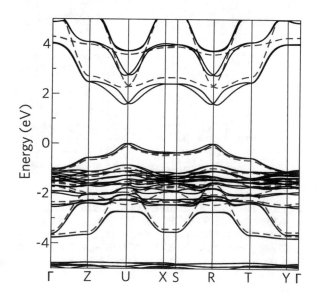

A. Ganose, *Atomic-Scale Insights into Emergent Photovoltaic Absorbers*,
Springer Theses, https://doi.org/10.1007/978-3-030-55708-9

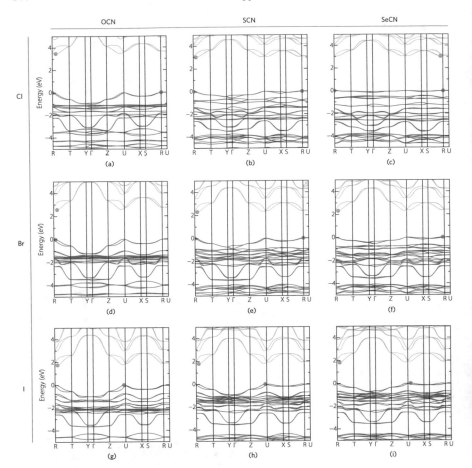

Fig. A.2 HSE06+SOC calculated band structures of lead-based MAPSI-structured analogues: **a** $(CH_3NH_3)_2Pb(OCN)_2Cl_2$, **b** $(CH_3NH_3)_2Pb(SCN)_2Cl_2$, **c** $(CH_3NH_3)_2Pb(SeCN)_2Cl_2$, **d** $(CH_3NH_3)_2Pb(OCN)_2Br_2$, **e** $(CH_3NH_3)_2Pb(SCN)_2Br_2$, **f** $(CH_3NH_3)_2Pb(SeCN)_2Br_2$, **g** $(CH_3NH_3)_2Pb(OCN)_2I_2$, **h** $(CH_3NH_3)_2Pb(SCN)_2I_2$, **i** $(CH_3NH_3)_2Pb(SeCN)_2I_2$. The valence band maximum is set to 0 eV. Valence and conduction bands indicated by blue and orange lines, respectively. Green and red circles indicate the valence band maximum and conduction band minimum, respectively

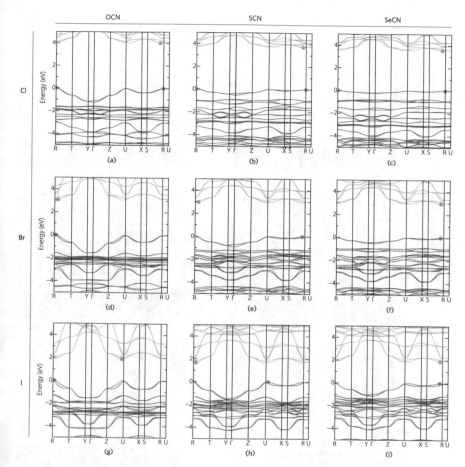

Fig. A.3 HSE06+SOC calculated band structures of tin-based MAPSI-structured analogues: **a** $(CH_3NH_3)_2Sn(OCN)_2Cl_2$, **b** $(CH_3NH_3)_2Sn(SCN)_2Cl_2$, **c** $(CH_3NH_3)_2Sn(SeCN)_2Cl_2$, **d** $(CH_3NH_3)_2Sn(OCN)_2Br_2$, **e** $(CH_3NH_3)_2Sn(SCN)_2Br_2$, **f** $(CH_3NH_3)_2Sn(SeCN)_2Br_2$, **g** $(CH_3NH_3)_2Sn(OCN)_2I_2$, **h** $(CH_3NH_3)_2Sn(SCN)_2I_2$, **i** $(CH_3NH_3)_2Sn(SeCN)_2I_2$. The valence band maximum is set to 0 eV. Valence and conduction bands indicated by blue and orange lines, respectively. Green and red circles indicate the valence band maximum and conduction band minimum, respectively

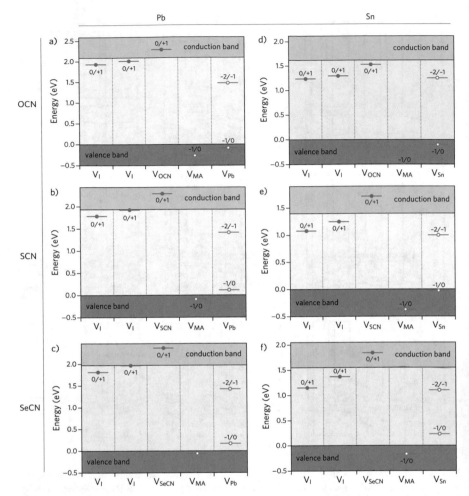

Fig. A.4 Charge-state transition level diagrams for: **a** $(CH_3NH_3)_2Pb(OCN)_2I_2$, **b** $(CH_3NH_3)_2Pb(SCN)_2I_2$, **c** $(CH_3NH_3)_2Pb(SeCN)_2I_2$, **d** $(CH_3NH_3)_2Sn(OCN)_2I_2$, **e** $(CH_3NH_3)_2Sn(SCN)_2I_2$, **f** $(CH_3NH_3)_2Sn(SeCN)_2I_2$, The valence band maximum is set to 0 eV. Valence and conduction bands indicated by blue and orange lines, respectively. Green and red circles indicate the valence band maximum and conduction band minimum, respectively. Red bands with filled circles indicate donor defects, green bands with open circles indicate acceptor defects

Table A.1 Relativistic lowering of the indirect ($\Delta_{\text{soc}} E_g^{\text{ind}}$) and direct ($\Delta_{\text{soc}} E_g^{\text{dir}}$) band gaps of $(CH_3NH_3)_2MPs_2I_2$, where M = Sn, Pb, and Ps = OCN, SCN, SeCN, calculated using HSE06 with and without spin–orbit coupling (SOC)

	Compound	$\Delta_{soc} E_g^{ind}$ (eV)	$\Delta_{soc} E_g^{dir}$ (eV)
Pb	$MA_2Pb(OCN)_2Cl_2$	0.75	0.82
	$MA_2Pb(SCN)_2Cl_2$	0.63	0.65
	$MA_2Pb(SeCN)_2Cl_2$	0.60	0.60
	$MA_2Pb(OCN)_2Br_2$	0.81	0.88
	$MA_2Pb(SeCN)_2Br_2$	0.70	0.73
	$MA_2Pb(OCN)_2I_2$	0.87	0.93
	$MA_2Pb(SCN)_2I_2$	0.77	0.78
	$MA_2Pb(SeCN)_2I_2$	0.73	0.76
Sn	$MA_2Sn(OCN)_2Cl_2$	0.16	0.15
	$MA_2Sn(SCN)_2Cl_2$	0.11	0.12
	$MA_2Sn(SeCN)_2Cl_2$	0.11	0.12
	$MA_2Sn(OCN)_2Br_2$	0.16	0.16
	$MA_2Sn(SCN)_2Br_2$	0.12	0.14
	$MA_2Sn(SeCN)_2Br_2$	0.12	0.14
	$MA_2Sn(OCN)_2I_2$	0.22	0.25
	$MA_2Sn(SCN)_2I_2$	0.20	0.23
	$MA_2Sn(SeCN)_2I_2$	0.20	0.22

Appendix B
Vacancy-Ordered Double Perovskites

See Fig. B.1.

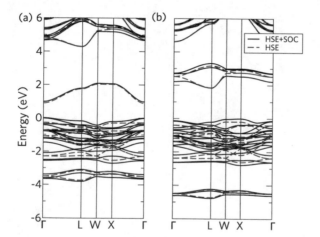

Fig. B.1 Effect of spin–orbit coupling on the band structures of **a** Cs_2SnI_6 and **b** Cs_2TeI_6. The HSE+SOC band structure is shown by black lines, with the HSE only band structure shown via dashed red lines. The valence band maxmimum of the HSE+SOC band structure is set to 0 eV

© The Editor(s) (if applicable) and The Author(s), under exclusive license
to Springer Nature Switzerland AG 2020
A. Ganose, *Atomic-Scale Insights into Emergent Photovoltaic Absorbers*,
Springer Theses, https://doi.org/10.1007/978-3-030-55708-9

Appendix C
Bismuth Chalcoiodides

See Fig. C.1.

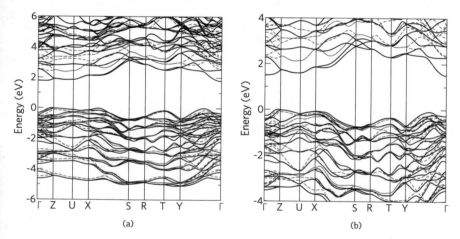

(a) (b)

Fig. C.1 Effect of spin–orbit coupling on the band structures of **a** BiSI and **b** BiSeI. The HSE+SOC band structure is shown by black lines, with the HSE only band structure shown via dashed red lines. The valence band maxmimum of the HSE+SOC band structure is set to 0 eV

A. Ganose, *Atomic-Scale Insights into Emergent Photovoltaic Absorbers*,
Springer Theses, https://doi.org/10.1007/978-3-030-55708-9

Printed in the United States
by Baker & Taylor Publisher Services